Studies in Infrastructure and Control

The book series aims to publish top-quality state-of-the-art textbooks, research monographs, edited volumes and selected conference proceedings related to infrastructure, innovation, control, and related fields. Additionally, established and emerging applications related to applied areas like smart cities, internet of things, machine learning, artificial intelligence, etc., are developed and utilized in an effort to demonstrate recent innovations in infrastructure and the possible implications of control theory therein. The study also includes areas like transportation infrastructure, building infrastructure management and seismic vibration control, and also spans a gamut of areas from renewable energy infrastructure like solar parks, wind farms, biomass power plants and related technologies, to the associated policies and related innovations and control methodologies involved.

Manish Yadav · Ashutosh K. Giri · Stepan Ozana

Control Strategies for Non-minimum Phase Systems

Springer

Manish Yadav
Department of Electrical Engineering
Sagar Institute of Science and Technology
Bhopal, Madhya Pradesh, India

Ashutosh K. Giri
Department of Electrical Engineering
Government Engineering College
Bharuch, Gujarat, India

Stepan Ozana
Department of Cybernetics and Biomedical
Engineering
Technical University of Ostrava
Ostrava, Czech Republic

ISSN 2730-6453 ISSN 2730-6461 (electronic)
Studies in Infrastructure and Control
ISBN 978-981-96-2350-1 ISBN 978-981-96-2351-8 (eBook)
https://doi.org/10.1007/978-981-96-2351-8

This Springer imprint is published by the registered company Springer Nature Singapore Pte Ltd.
The registered company address is: 152 Beach Road, #21-01/04 Gateway East, Singapore 189721, Singapore

Preface

Control Systems Engineering combines the thrill of problem-solving with the challenge of integrating diverse disciplines. This book is designed and organized around the concepts of control systems engineering for Non-Minimum Phase (NMP) zeros and dead-time systems. Chapter 1 presents a brief outline of Non-Minimum Phase system. An overview of the primary strategy for controlling physical variables in systems is provided, including definitions of frequently used terms related to control system operation, analysis, and design. Chapter 1 is started with the basic idea of inverse response. Chapter 2 presents an improved series cascade control scheme with an inverse response compensator for non-minimum phase with dead-time systems. Chapter 3 displays an enhanced series cascade control scheme with an inverse response compensator for non-minimum phase with dead-time systems. Chapter 4 discusses an improved structure of a parallel cascade scheme with different filters for the non-minimum phase systems. Chapter 5 is devoted modified tuning approach with optimal fractional order IMC-based series cascade control for the non-minimum phase systems. Chapter 6 is devoted to tuning of fractional order filter with non-minimum phase zeros. Chapter 7 indicates a combined approach for tuning the fractional IMC filter with a series cascade scheme for the non-minimum phase systems. Finally, Chap. 8 devotes the conclusions and future works. Our aim with this book is to cultivate a genuine interest in control systems analysis and design among students. For those who wish to explore the topic further, an extensive bibliography is available at the end of the book of each chapter.

Bhopal, India — Manish Yadav
Bharuch, India — Ashutosh K. Giri
Ostrava, Czech Republic — Stepan Ozana

Contents

1 Outline of Non-minimum Phase Systems 1
1.1 Preliminaries 1
1.2 Problem Statement and Basics 2
1.3 Fractional-Order Controller in a Single Loop 3
1.4 Cascade Control Strategies 3
1.5 Inverse Response Compensator 4
1.6 Stimulus 5
1.7 Modification of Series Cascade Control Design 7
1.8 Modification of Parallel Cascade Control Scheme 7
1.9 Some Informative Views of Chapter in Post-Mortem Form 7
References 12

2 Refined Series Cascade Control Techniques for Handling Non-minimum Phase Zeros and Dead Time in Control Systems 15
2.1 A Refined Series Cascade Control Architecture 15
2.1.1 Conceptual Evolution 16
2.1.2 Series Cascade Control with IMC Approach 16
2.1.3 Fractional IMC Controller with Inverse Response and Dead-Time Compensator 17
2.1.4 Case Studies 22
2.2 Decomposed Series Cascade Control Scheme 32
2.2.1 Case Studies 37
2.3 Improved Control Structure with Noise Filter 44
2.3.1 Modified Series Cascade Control with Noise Filter 44
2.3.2 Noise Filter Design 44
2.3.3 Case Studies 45
2.4 Summary 47
2.5 Some Informative Views of Chapter in Post-Mortem Form 47
References 52

3 Advancements in the Cascade Control Scheme 55
3.1 An Improvement in a Series Cascade Control Scheme 55
3.2 Series Cascade Control Scheme with Inverse Response Compensator 55
3.3 Case Studies 60
3.3.1 Case Study I 60
3.3.2 Case Study II 62
3.3.3 Case Study III 67
3.3.4 Case Study IV 69
3.4 Control Structure with Derivative Filter 73
3.5 Controller Design 75
3.6 Case Studies 79
3.6.1 Case Study I 80
3.6.2 Case Study II 85
3.7 Summary 90
3.8 Some Informative Views of Chapter in Post-mortem Form 90
References 93

4 Improved Parallel Cascade Control for Non-minimum Phase System 95
4.1 An Enhancement in Parallel Cascade Scheme 95
4.2 Parallel Cascade Scheme 95
4.3 Controller Design 97
4.4 Case Studies 99
4.4.1 Case Study I 100
4.4.2 Case Study II 105
4.4.3 Case Study III 109
4.4.4 Case Study IV 112
4.5 Summary 117
4.6 Some Informative Views of Chapter in Post-mortem Form 117
References 120

5 Modified Tuning Approaches for Non-minimum Phase Systems 123
5.1 Introduction 123
5.2 Modified Tuning Approach for Fractional-Order Filter 123
5.3 Theoretical Development 124
5.4 Delayed Bode's Ideal Transfer Function Approach 124
5.5 Suggested Work 125
5.6 Particle Swarm Optimization (PSO) 127
5.7 Case Studies 128
5.7.1 Case Study I 128
5.7.2 Case Study II 132
5.8 Summary 136
5.9 Some Informative Views of Chapter in Post-mortem Form 137
References 143

6 Tuning of Fractional Order Filter with Non-minimum Phase Zero 145
6.1 An Improved Series Cascade Control Structure 145
6.2 Controller Design 147
6.3 Search Space Reduction (SSR) Algorithm 149
6.4 Case Studies 151
6.4.1 Case Study I 151
6.4.2 Case Study II 155
6.5 Summary 158
6.6 Some Informative Views of Chapter in Post-mortem Form 159
References 160

7 A Combined Approach for Tuning Optimal Fractional IMC-Based Series Cascade Control 163
7.1 Introduction 163
7.2 Suggested Work 163
7.3 Particle Swarm Optimization (PSO) 168
7.4 Case Studies 168
7.4.1 Case Study I 168
7.4.2 Case Study II 172
7.5 Summary 174
7.6 Some Informative Views of Chapter in Post-mortem Form 176
References 177

8 Conclusion and Future Scope 179
8.1 Summary of Findings 179
8.2 Future Scope 180

About the Authors

Dr. Manish Yadav is an Associate Professor in the Electrical Engineering Department at Sagar Institute of Science and Technology, Bhopal. He was a research scholar in the Department of Electrical Engineering at S. V. National Institute of Technology, Gujarat, India. He received his B.E. degree from Maharana Pratap College of Technology, Gwalior, Madhya Pradesh, India, in 2006 and Master's degree in Measurement and Control from the Madhav Institute of Technology, Gwalior, Madhya Pradesh, India, in 2011. He received Ph.D. degree from S. V. National Institute of Technology, Gujarat, India, in 2023. His area of research is fractional order controller design. He has published seven research articles and one book chapter.

Dr. Ashutosh K. Giri received his Bachelor of Engineering in Electrical Engineering from College of Technology, G. B. Pant University of Agriculture and Technology, Pantnagar Uttarakhand, India, in 2002, M.E. in Electrical Engineering with specialization in Power System from the L. D. College of Engineering, Ahmedabad, Gujarat, India, in 2005, and Ph.D. from the Department of Electrical Engineering, Sardar Vallabh National Institute of Technology, Surat, India, in 2020. In June 2005, he joined the Department of Electrical Engineering, PIET Limda (now deemed university) Vadodara as a lecturer. In May 2011, he joined the Department of Electrical Engineering, Government Engineering College, Bharuch, as an assistant professor. He has authored and co-authored more than 75 research papers in reputed journals such as IEEE Transactions, IET Journal, Wiley, Tailor Francis, and Springer.

Dr. Stepan Ozana was born in Bilovec, Czech Republic, in May 1977. He studied electrical engineering at the V. S. B. Technical University of Ostrava and received the M.Sc. degree in control and measurement engineering, in 2000, and the Ph.D. degree in technical cybernetics, in 2004. In 2015, he was habilitated in technical cybernetics. Since then, he has been working as an associate professor with the Department of Cybernetics and Biomedical Engineering, Faculty of Electrical Engineering and Computer Science, V. S. B. Technical University of Ostrava. He currently gives lectures on cybernetics and control systems. His main areas of interest and expertise are modeling and simulation of dynamic systems, control theory, automation, design, implementation, and deployment of control algorithms using softPLC systems.

Chapter 1
Outline of Non-minimum Phase Systems

"Nothing is impossible. We merely don't know how to do it yet."
—Thomas A. Edison

1.1 Preliminaries

The inverse response occurs when initially opposing effects within a system result in an initial reaction that differs from the eventual steady-state outcome. This trait, frequently observed in non-minimum phase systems, arises from conflicting dynamics [1]. Such phenomena are prevalent in industries such as process control, for example, in boiler drum level systems [2], chemical reactors such as the van de Vusse reactor [3], or the re-boiler section of distillation columns. The occurrence of inverse response can typically be attributed to Right Half Plane (RHP) open-loop zeros.

For instance, in a boiler drum level control scenario, an increase in the cold feed-water flow rate initially decreases the volume of boiling water, thereby momentarily reducing the liquid level, before the level eventually starts to rise. From Fig. 1.1, the outcome reflects the interplay of the conflicting effects,

$$\frac{K_2'}{s} - \frac{K_1'}{\tau_1' s + 1} = \frac{(K_2'\tau_1' - K_1')s + K_2'}{s(\tau_1' s + 1)} \tag{1.1}$$

From Eq. (1.1), we notice that the transfer function has a positive zero at the point $s = -K_2' / (K_2'\tau_1' - K_1')$. The interaction of two conflicting effects explains the emergence of the inverse response in this example (Fig. 1.2).

M. Yadav et al., *Control Strategies for Non-minimum Phase Systems*,
Studies in Infrastructure and Control, https://doi.org/10.1007/978-981-96-2351-8_1

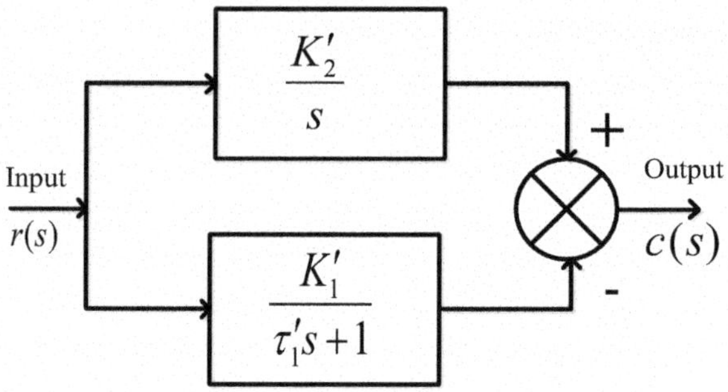

Fig. 1.1 Liquid level in a boiler system

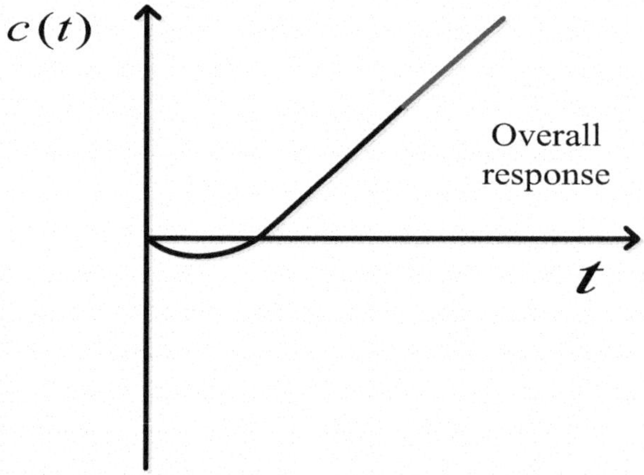

Fig. 1.2 Inverse response behavior

1.2 Problem Statement and Basics

The NMP characteristic complicates achieving a practical closed-loop response due to the challenging dynamics it introduces for automatic controllers. As a result, significant constraints in achieving the desired performance emerge, and it becomes impossible to achieve perfect control with any stable causal controller. The key limitations linked to the presence of RHP zeros include issues such as high-gain instability and bandwidth constraints. These factors contribute to the sluggish response observed in processes affected by inverse response [4]. Moreover, the closed-loop stability of such systems is inherently at risk. A reliable source can find discussions on PID control in non-minimum phase systems within specialized control theory publications [5–7].

1.3 Fractional-Order Controller in a Single Loop

A fractional-order controller is an advanced type of control system where the order of the controller's differential or integral actions is not limited to integer values. Instead, it can take fractional values, which allows for greater flexibility and potentially improved performance in certain systems. Here's a brief overview of how a fractional-order controller operates in a single-loop system:

Concept of Fractional-Order Control

- ***Fractional-Order Derivative and Integral:*** Unlike traditional controllers that use integer-order derivatives and integrals, fractional-order controllers use derivatives and integrals of arbitrary real (fractional) orders.

1.4 Cascade Control Strategies

Effective performance and quick response are crucial requirements for implementing control systems in industrial environments. PID controllers are widely adopted in industry for their straightforward system modeling and their historical evolution through practical tuning methods over the past few decades [5]. Even in more complex control scenarios, PID controllers serve as the foundation upon which more sophisticated control strategies can be built, especially when integrating multiple simple control systems.

Cascade control systems prove particularly effective when single-loop PID controllers face challenges in sufficiently handling disturbances. The core principle of cascade control involves minimizing the influence of disturbances on the primary controlled variable by utilizing an internal loop (or inner loop) where an intermediate measurement is accessible [6].

Cascade control setups can be classified as either Series or Parallel types. In a series cascade control system, disturbances that affect the intermediate variable are swiftly addressed by the inner loop. This preemptive action reduces the impact of these disturbances on the outer loop, leading to better overall disturbance management y_1. By handling fluctuations in the intermediate variable through the inner loop, the outer loop experiences fewer disturbances, which helps to minimize the error in controlling the final process variable. The disturbances $G_{d_2}(s)$ are mitigated by the inner loop controller earlier they disrupt the main process, resulting in quicker consequence times and minimizing the influence to y_1 is minimum (Fig. 1.3).

Figure 1.4 demonstrates that in parallel cascade control, the manipulated variable and disturbances concurrently impact the performance of both the outer and inner loops. In Fig. 1.4, $P_1(s)$ and $P_2(s)$ are the transfer functions of the outer and inner loop processes respectively. d_2 and d_p are the transfer functions of the disturbances for inner and outer loops correspondingly [7]. By effectively attenuating disturbances within the inner loop, it minimizes steady-state errors in the outer loop. This

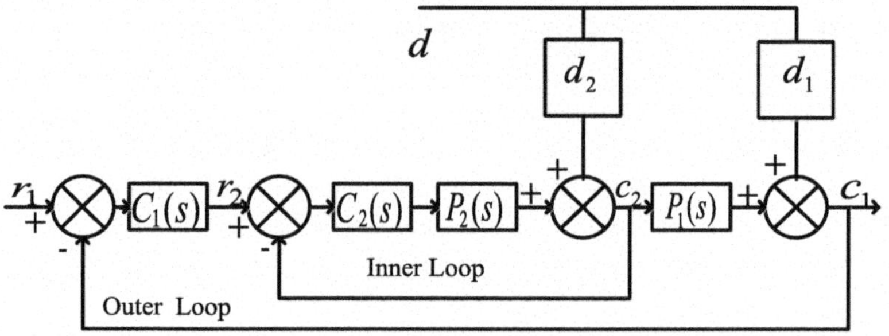

Fig. 1.3 Series cascade control scheme

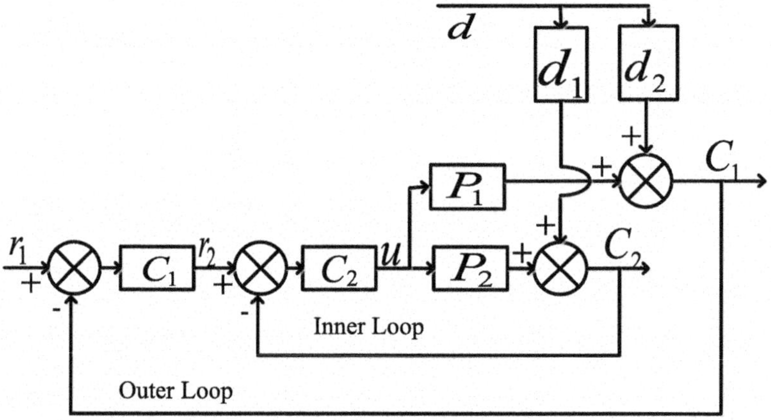

Fig. 1.4 Parallel cascade control scheme

approach proves particularly beneficial in scenarios where outer loop measurements are infrequent or subject to significant delays.

Once the inner loop is properly tuned, attention shifts to the outer loop controller, incorporating the optimized inner loop settings. This method is more elaborate and time-consuming compared to standard single-loop PID control techniques. Contributions from various researchers have advanced the development of series cascade control [8] and parallel cascade control approaches [9, 10].

1.5 Inverse Response Compensator

The inverse response compensator (Fig. 1.5) in feedback configurations is widely appreciated for its utility. Its key advantage is that it allows for the use of design methodologies that are generally applied to systems. This approach integrates a

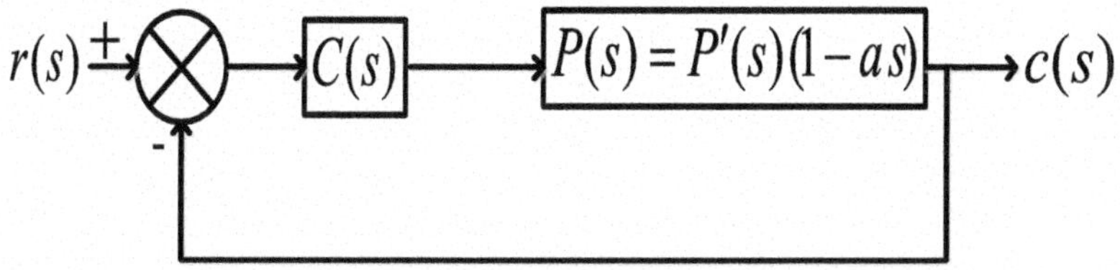

Fig. 1.5 Traditional controller with inverse response compensator

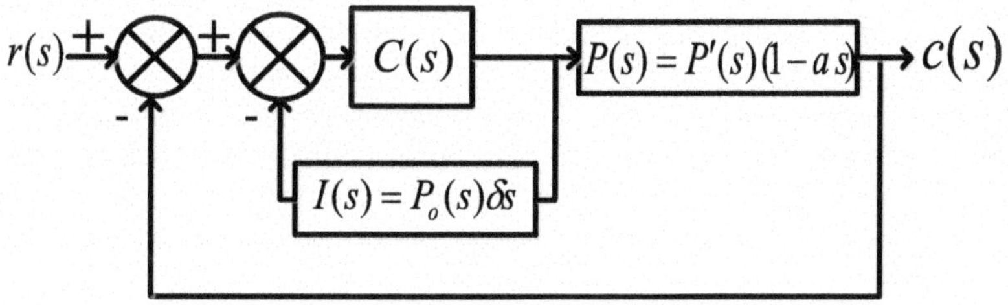

Fig. 1.6 Schematic diagram incorporating inverse response compensator

process's mathematical model within an internal feedback loop with a conventional controller [11].

Referring to Fig. 1.6, you can see that a minor loop is incorporated into the conventional controller transfer function.

If we select δ such, the counteractive signal from the minor loop successfully neutralizes the inverse response in the feedback loop. Conversely, care must be taken not to choose a value that is too large. In this book, it is selected,

$$\delta = 2a.$$

1.6 Stimulus

In today's rapidly evolving global competitive environment, control engineers encounter increasingly demanding challenges, including stringent regulations and intricate integrated processes. Many existing control theories and design approaches assume predefined specifications for the control system, often overlooking the significant complexities involved in actual system design. In practice, control theory and real-world process control applications often diverge due to unforeseen external disturbances, variations in plant parameters, and unaccounted-for dynamics. Industrial systems exhibit a wide range of characteristics, from Single-Input Single-Output (SISO) to Multi-Input Multi-Output configurations, with features such as dead time, non-minimum phase behavior, instability, and higher-order dynamics. The primary goal of this book is to introduce new theories and methodologies to bridge this gap. An effective control system should respond swiftly with minimal overshoot and demonstrate robustness in the face of process uncertainties. Achieving optimal performance requires well-tuned process controllers tailored to the specific characteristics of diverse industrial processes. Consequently, this book explores different control architectures and methodologies.

PID control remains a widely adopted control approach due to its straightforward implementation and effectiveness [5]. However, the versatility of PID controllers extends beyond integer orders; they can also operate effectively at fractional orders. [12] introduced the concept of fractional-order controllers, which offer an additional degree of freedom for tuning. This book addresses the challenges associated with designing robust controllers for systems characterized by uncertain time delays and non-minimum phase behaviors by exploring advanced fractional-order control techniques. These controllers leverage fractional calculus principles to enhance performance in complex dynamic systems, providing a more nuanced approach to achieving robust and stable control.

PID controllers demonstrate effective performance for linear processes with non-minimum phase (NMP) zeros. However, in process industries, certain systems exhibit initial responses in the opposite direction due to NMP zeros, posing challenges that cannot be adequately addressed by a standard PID controller. This limitation arises because the phase angle associated with the NMP zero restricts the controller's effectiveness. Many control strategies prioritize setpoint tracking while neglecting the crucial aspect of disturbance rejection performance.

Conversely, in industrial control applications, the ability to reject disturbances often outweighs the importance of precise setpoint tracking [5]. Setpoint references may remain unchanged for extended periods, while the system's performance is predominantly influenced by varying disturbances. Effective management of disturbances is crucial for both productive and inefficient applications. Although a single controller can manage disturbances in a feedback system, there is usually a trade-off between improving setpoint tracking and rejecting disturbances. Non-minimum phase (NMP) zeros and time delays in systems often result in significant overshoot, undershoot, and extended settling times, underscoring the complexities of control design. An advanced approach involves introducing an additional controller specifically to enhance disturbance rejection while preserving the system's stability and minimizing undesired responses. This strategy aims to significantly improve disturbance rejection capabilities without compromising on the benefits of existing control methods, even under conditions of model uncertainties [13–15]. Controlling systems with significant time delays and NMP characteristics remains a challenging area in control engineering. The critical nature of controller operation in non-minimum phase systems underscores ongoing research efforts.

Cascade control is a methodology employed to improve disturbance rejection performance across various applications [8]. The fundamental principle of cascade control involves minimizing the impact of disturbances on the outer loop-controlled variable by employing an inner (or secondary) loop that utilizes intermediate measurement disturbances in industrial processes.

1.7 Modification of Series Cascade Control Design

In a series cascade configuration, the adapted outer loop controller integrates a fractional filter, an inverse response compensator, and a dead-time compensator to manage non-minimum phase (NMP) zeros in systems with dead-time processes. The fractional filter enhances control adaptability, the inverse response compensator addresses inverse process behaviors, and the dead-time compensator reduces the initial impacts of time delays in the feedback loop.

1.8 Modification of Parallel Cascade Control Scheme

The book introduces a sophisticated framework for the outer loop controller in a parallel cascade setup, which combines a fractional filter, an inverse response compensator, and a dead-time compensator to deal with non-minimum phase (NMP) zeros in systems that experience dead time. This methodology presents a unified theory to optimize the outer loop controller strategy, using advanced fractional filters and resilient compensators for both dead time and inverse response issues. It aims to minimize setpoint fluctuations and employs an Internal Model Control (IMC) approach to streamline the development process.

1.9 Some Informative Views of Chapter in Post-Mortem Form

Question 1: What impact does dead time have on the response of basic feedback control loops?

Answer: *Dead time in simple feedback control loops can have significant effects on their response characteristics.*

Key effects of dead time in feedback control loops include.

1. *Delayed Response*
2. *Instability Risk*
3. *Reduced Control Precision*
4. *Difficulty in Tuning*
5. *Impact on System Dynamics*

Question 2: Describe the physical terms of the notion of dead-time compensation. Why such a system is also called a predictor?

Answer: *Dead-time compensation in control systems refers to the techniques and methods used to mitigate the effects of dead time, which is the delay between an* *action or disturbance in a system and its measurable effect on the output. In physical terms, dead time arises due to factors such as transport delays in fluid systems, communication delays in networks, or processing delays in electronic systems.*

Here's how dead-time compensation works in physical terms.

1. *Prediction Mechanism*
2. *Model-Based Approaches*
3. *Feedforward Control*

The term "predictor" is often used because dead-time compensation systems essentially calculate the future state of the system's output based on current and past observations. By predicting how the system will respond to inputs despite the dead time, these systems "predict" or estimate the future behavior that would occur if there were no delay. This predictive capability allows the controller to proactively adjust its actions to maintain stability, improve response times, and enhance overall control performance.

Question 3: What is the impact of model inaccuracies on the efficacy of dead-time compensator?

Answer: *The model inaccuracies can compromise the usefulness of a dead-time compensator by reducing prediction accuracy, causing stability issues, degrading performance, decreasing robustness, and necessitating more advanced control strategies to mitigate these effects. Therefore, careful model identification and validation are crucial in designing effective dead-time compensators for control systems.*

Questions 4: What is our goal when designing a controller for a system with an inverse response compensator?

Answer: *When crafting a controller for a system exhibiting inverse response, the primary objective is to ensure stability, mitigate or reduce the impacts of inverse response, enhance response speed, bolster robustness, achieve optimal control performance, and maintain.*

Question 5: What are the key benefits and shortcomings of a cascade control scheme? For what kind of processes can you employ cascade control?

Advantages of Cascade Control:

Answer: *Flexibility: Cascade control is flexible and can be applied to processes with complex dynamics or interacting variables. It allows for more precise control of both primary and secondary variables independently.*

Reduced Offset: In some cases, cascade control can help reduce offset (steady-state error) compared to single-loop control systems. This is especially beneficial in processes where achieving precise setpoint control is crucial.

Disadvantages of Cascade Control:

Design and Tuning Challenges: Proper design and tuning of cascade control systems require expertise and a detailed understanding of the process dynamics. Incorrect tuning can lead to instability or suboptimal performance.

Increased Cost: Implementing cascade control may require additional sensors, actuators, and controllers, which can increase system costs compared to simpler control schemes.

Maintenance and Diagnosis: Maintenance and troubleshooting of cascade control systems can be more challenging due to their complexity. Identifying and resolving issues requires comprehensive knowledge of the entire control structure.

Processes Suitable for Cascade Control:

Cascade control is particularly suitable for processes where:

Slow Secondary Dynamics: The secondary variable responds more slowly to disturbances compared to the primary variable. Examples include temperature control in reactors where the jacket temperature (secondary) affects the reactor temperature (primary).

Interacting Variables: Processes where the primary and secondary variables interact and influence each other. For instance, flow and pressure control in hydraulic systems where flow rate (primary) depends on pressure (secondary).

Complex Dynamics: Processes with complex dynamics where direct control of the primary variable alone is insufficient to achieve desired performance. Cascade control allows for better handling of these dynamics by separating control tasks.

Critical Processes: Processes where precise control of both primary and secondary variables is essential for maintaining product quality, safety, or efficiency.

Question 6: What type of controllers would you use for the two controllers of a cascade system? How would you tune them?

Answer: *In a cascade control system, you typically use two controllers: an* **outer (primary) controller** *and an* **inner (secondary) controller**. *Each controller serves a specific role in controlling the process variables. Here's how you would select and tune these controllers*:

Outer (Primary) Controller:

- ***Type of Controller***
- ***Tuning***: *The tuning process typically involves*:
  - ***Initial Setting***: *Start with Kp and Ki values and set Kd to zero.*
  - ***Adjusting Kp***: *Increase until the system responds quickly without significant overshoot.*
  - ***Adjusting Ki***: *Increase to eliminate steady-state error without causing instability.*
  - ***Adding Kd***: *Fine-tune to dampen oscillations and improve stability.*

Advanced tuning methods such as Ziegler-Nichols or iterative methods can be used to expedite the tuning process.

Inner (Secondary) Controller:

- ***Type of Controller***
- ***Tuning***: *Tuning the inner controller follows a similar process to the outer controller but is typically faster and more responsive*:
  - ***Proportional Gain*** (K_p): *Adjust to achieve the desired response speed without causing instability.*
  - ***Integral Gain*** (K_i): *Fine-tune to eliminate any steady-state error in controlling the secondary variable.*
  - ***Derivative Gain*** (K_d): *If needed, add to improve damping and reduce overshoot.*

Integration and Tuning Considerations:

- ***Interaction Effects***: *Cascade control systems require careful consideration of interaction effects between the primary and secondary loops. Ensuring stability and optimal performance often involves iterative tuning and testing to balance the response of both controllers.*
- ***Simulation and Validation***: *Before implementing cascade control on the actual process, simulation tools can be used to validate controller settings and assess system performance under various scenarios.*

Question 7: Why do you think a system with inverse response is difficult to control?

Answer: *A system with an inverse response is challenging to control primarily due to its unusual and counterintuitive dynamics. Here's why*:

1. ***Non-intuitive Behavior***: *An inverse response system initially moves in the opposite direction of the desired control action before eventually moving in the intended direction. This counterintuitive behavior makes it difficult to predict and compensate for the system's behavior in real time.*
2. ***Initial Disturbance***: *When a control input is applied, the system might exhibit an initial response in the opposite direction of the input. This can cause instability if not properly accounted for in the control design, as the initial negative response might lead to overshooting or oscillations before settling into the desired behavior.*
3. ***Increased Tuning Difficulty***: *Tuning a controller for a system with inverse response involves adjusting parameters to handle both the initial inverse behavior and the eventual desired response. This process can be complex and requires careful consideration to avoid instability and ensure that the system performs well under various conditions.*
4. ***Potential for Oscillations***: *The inverse response can introduce oscillations or instability if not properly managed. Since the system initially moves in the opposite direction, it can lead to a situation where the controller's efforts to correct*

the response can exacerbate the initial inverse behavior, resulting in oscillatory or unstable performance.

Question 8: What is an inverse response and what causes it?

Answer: *An inverse response in a control system refers to a situation where the system's output initially moves in the opposite direction of the desired input before eventually moving in the intended direction. This counterintuitive behavior can be challenging to control and predict.*

Causes of Inverse Response

1. ***System Dynamics***:
   - ***Time Delays***: *If a system has significant time delays in its dynamics, it might show an inverse response due to the delay in the system's reaction. The system's output might initially move in the opposite direction before the delay is overcome and the system starts moving in the intended direction.*
   - ***Zero Dynamics***: *Certain systems feature zero dynamics, which are internal modes that can display inverse behavior.*
2. ***Nonlinearities***:
   - ***Dead Time***: *Systems with dead time (a period during which the system output does not change in response to an input) can exhibit inverse responses. The dead time can cause the system to temporarily react in the opposite direction before settling into the desired direction.*
   - ***Saturation Effects***: *If the system experiences nonlinear effects such as saturation, where the output cannot increase or decrease beyond a certain limit, the initial effect of an input change might be counter to the intended effect until the system adjusts.*
3. ***Feedback Loop Interactions***:
   - ***Feedback Configuration***: *In some feedback control systems, particularly those with complex or poorly designed feedback loops, Initially, the feedback dynamics may make the system move in the opposite direction of the input change, but the feedback loop will eventually correct this and steer the system as intended.*
4. ***Mechanical Systems***:
   - ***Inertia and Dynamics***: *In mechanical systems, particularly those involving actuators with significant inertia, the system may exhibit an inverse response due to the inertial effects. For instance, if an actuator is initially displaced in the opposite direction due to its own dynamics before the desired motion is achieved, it can create an inverse response.*

Understanding and designing for inverse responses require careful analysis of the system dynamics and often involve advanced control techniques to manage the initial counterintuitive behavior and ensure the system eventually behaves as desired.

Question 9: What is a fractional-order controller, and how does it differ from a traditional integer-order controller?

Answer: *Fractional-order controllers offer greater flexibility and can be tuned to achieve better performance in systems with specific dynamic characteristics. They can improve robustness, reduce overshoot, and enhance stability. Additionally, fractional-order controllers can model processes more accurately when the system dynamics are not well captured by integer-order PID controller dynamics.*

References

1. Vidyasagar M (1986) On undershoot and non-minimum phase zeros. IEEE Trans Autom Control 31(5):440–440
2. Pai NS, Chang SC, Huang CT (2010) Tuning PI/PID controllers for integrating processes with dead-time and inverse response by simple calculations. J Process Control 20(6):726–733
3. Luyben WL (2000) Tuning proportional-integral controllers for processes with both inverse response and dead time. Ind Eng Chem Res 39(4):973–976
4. Iionya K, Alpeter RJ (1962) Inverse response in process control. Ind Eng Chem 54(7):39–43
5. Åström KJ, Murray RM (2008) Feedback systems: an introduction for scientists and engineers. Princeton University Press, New Jersey
6. Raja GL, Ali A, Series cascade control an outline survey. In: Indian control conference. Guwahati, India, pp 409–414
7. Padahn DG, Manjhi S (2012) Modified smith predictor-based cascade control of unstable time delay processes. ISA Trans 51(1):95–104
8. Franks RG, Worley CW (1956) Quantitive analysis of cascade control. Ind Eng Chem 48(6):1074–1079
9. Rao AS, Seethaladevi S, Uma S, Chidambaram M (2009) Enhancing the performance of parallel cascade control using smith predictor. ISA Trans 48:220–227
10. Pashaei S, Bagher P (2020) Parallel cascade control of dead time process via fractional order controller based on smith predictor. ISA Trans 98:186–197
11. Stephanopoulos G (1984) Chemical process control. Englewood Cliffs, New Jersey, Prentice Hall
12. Podlubny I (1999) Fractional-order systems and $PI^{\lambda}D^{\mu}$ controllers. IEEE Trans Autom Control 44(1):208–214
13. Nagarsheth SH, Sharma SN (2020) Control of non-minimum phase system with dead time: a fractional system viewpoint. Int J Syst Sci 51(11):1905–1928
14. Singh AP, Deb D, Agarwal H (2019) On selection of improved fractional model and control of different systems with experimental validation. Commun Nonlinear Sci Numer Simul (Elsevier) 79:10490
15. Rajkumar SM, Chakraborty S, Dey R, Deb D (2020) Online delay estimation and adaptive compensation in wireless networked system: an embedded control design. Int J Control Autom Syst 18:856–866
16. Joshi D, Deb D, Giri AK (2024) MRAS disturbance observer-based sensorless field-oriented backstepping control of BLDC motor drive. In: Electrical engineering. Springer, Berlin. https://doi.org/10.1007/s00202-024-02378

17. Kundu S, Singh M, Giri AK (2022) Implementation of in-loop filter based DSC-FLL control approach for power quality improvement in isolated microgrid. In: Proceedings of 2022 IEEE 31st international symposium on industrial electronics (ISIE). USA. https://doi.org/10.1109/ISIE51582.2022.9831712
18. Giri AK, Arya SR, Maurya R (2020) Hybrid order generalized integrator based control for VSC to improve the PMSG operation in isolated mode. In: The proceedings of 2020 first international conference on power, control and computing technologies (ICPC2T). India. https://doi.org/10.1109/ICPC2T48082.2020.9071498
19. Kundu S, Singh M, Giri AK (2024) Synchronization and control of WECS-SPV-BSS-based distributed generation system using ICCF-PLL control approach. Electr Power Syst Res 226:109919
20. Giri AK, Arya SR, Guerrero JM, Kumar S, Power quality: infrastructures and control. Springer Nature, Singapore. ISBN: 978-981-19-7956-9
21. Giri AK, Arya SR, Vinnikov D, Distributed energy systems design, modeling, and control, vol 1. CRC Press, pp 1–308. ISBN: 9781032134246
22. Arya SR, Patel MM, Alam SJ, Srikakolapu J, Giri AK, Chitti Babu B (2021) Classical control algorithms for permanent magnet synchronous generator driven by diesel engine for power quality. Int J Circuit Theory Appl 49(3):576–601
23. Kundu S, Giri AK, Singh M, Kadiyan S, Adaptive control strategy for isolated renewable energy-based generation system with intermittency. In: Electrical engineering. Springer, Berlin. https://doi.org/10.1007/s00202-024-02454-0
24. Meena DC, Singh M, Giri AK (2021) Leaky-momentum control algorithm for voltage and frequency control of three-phase SEIG feeding isolated load. J Eng Res ICARI Spec 109:120
25. Meena DC, Singh M, Giri AK (2021) A modified NLMS control algorithm for coordinated operation in three-phase wind-energy conversion system. Int J Renew Energy Res (IJRER) 11(4):1621–1629
26. Dalwadi N, Deb D, Ozana S (2023) Performance evaluation of RW-quadrotor and Bi-quadrotor for payload delivery. IET Intel Transp Syst 17(11):2221–2236
27. Dalwadi N, Deb D, Ozana S, Adaptive hybrid control of quadrotor drones. 461:1–174. ISBN: 978-981-19-9743-3

Chapter 2
Refined Series Cascade Control Techniques for Handling Non-minimum Phase Zeros and Dead Time in Control Systems

The NMP zero and dead-time systems are pervasive in process industries, significantly complicating controller design. Furthermore, when noise affects the system, it can lead to undesirable control actions, causing premature wear of actuators and undermining closed-loop performance. Using lower-order filters offers a compromise between setpoint tracking and robustness. To address these challenges, this chapter explores: (i) refining the series cascade control structure by incorporating fractional filters, inverse responses, and dead-time compensators within the IMC framework, (ii) improving the series cascade approach with fractional filters, inverse responses, dead-time compensation, and noise filters. (iii) Decomposed series cascade control scheme.

2.1 A Refined Series Cascade Control Architecture

Series cascade control is comprehensively utilized in process industries, where outer loop outcomes are influenced by manipulated variables and disturbances from inner loop outputs [1]. Recent surveys and modifications in series cascade control have been introduced [2]. In the literature, time delay terms are often approximated using Taylor series or Pade approximations for controller design purposes [3, 4]. However, no integrated theory exists that comprehensively designs outer loop controllers for series cascade control, encompassing fractional filters, inverse response compensators, and dead-time compensators with diverse approximations. Further, the Kalman filter and variable gain controller are presented [5–7].

M. Yadav et al., *Control Strategies for Non-minimum Phase Systems*,
Studies in Infrastructure and Control, https://doi.org/10.1007/978-981-96-2351-8_2

2.1.1 Conceptual Evolution

In process industries using cascade control, the inner loop is typically characterized by stable dynamics, whereas the outer loop can present various forms of dynamic behavior such as stability, non-minimum phase, integration, or instability.

2.1.2 Series Cascade Control with IMC Approach

The transfer function of the inner and outer loop of series cascade control becomes,

$$\frac{c_2(s)}{r_2(s)} = \tilde{P}_2(s)C_2(s), \tag{2.1}$$

$$\frac{c_2(s)}{d_2(s)} = 1 - \tilde{P}_2(s)C_2(s), \tag{2.2}$$

$$\frac{c_1(s)}{d_1(s)} = \frac{(1 - \tilde{P}_1(s)P_1(s)C_1(s))d_1(s) + (1 - \tilde{P}_2(s)C_2(s))P_1(s)d_2(s)}{1 + (P_1(s) - \tilde{P}_1(s))P_2(s)C_1(s) + (P_2(s) - \tilde{P}_2(s)C_2(s))}. \tag{2.3}$$

$$\frac{c_1(s)}{r_1(s)} = \frac{\tilde{P}_1(s)\tilde{P}_2(s)C_1(s)C_2(s)}{1 + (P_1(s) - \tilde{P}_1(s))P_2(s)C_1(s) + (P_2(s) - \tilde{P}_2(s)C_2(s)}, \tag{2.4}$$

In Fig. 2.1, the controllers, $C_1(s)$, $C_2(s)$ are termed master and slave controllers, respectively. This generalized arrangement consists of two plants $P_1(s)$,$P_2(s)$. In the IMC framework, both plants have their process models $\tilde{P}_1(s)$. In this generalized scheme (Fig. 2.1), $r_1(s)$, $c_1(s)$, $d_1(s)$, and $r_2(s)$, $c_2(s)$, $d_2(s)$ are the input, output, and disturbance for the outer and inner loop [8].

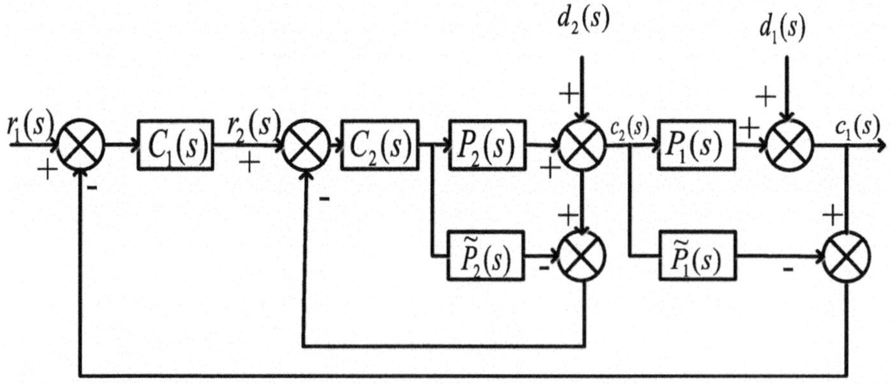

Fig. 2.1 Series cascade control with IMC scheme

2.1.3 Fractional IMC Controller with Inverse Response and Dead-Time Compensator

In Eqs. (2.5) and (2.6), the proposed control strategy is demonstrated (Fig. 2.2). After embedding both compensators, the outer loop controller can compensate for inverse response behavior and the large dead time of the process

$$\frac{c_1(s)}{r_1(s)} = \frac{P_1(s)P_2(s)C_2(s)\{C_1(s)(I_0(s)+D_0(s))\}}{1+(P_1(s)-\tilde{P}_1(s))P_2(s)\{C_1(s)(I_0(s)+D_0(s)\}+P_2(s)-\tilde{P}_2(s)C_2(s))}. \tag{2.5}$$

$$\frac{c_{1(s)}}{d_1(s)} = \frac{(1-\tilde{P}_1(s)P_1(s)C_1(s)(I_o(s)+D_o(s))+(1-\tilde{P}_2(s)C_2(s))P_1(s)d_2}{1+(P_1(s)-\tilde{P}_1(s))P_2(s)\{C_1(s)(I_o(s)+D_o(s))+P_2(s)-\tilde{P}_2(s)C_2(s)}. \tag{2.6}$$

Inner loop controller design: The IMC framework decomposes the process model into two distinct segments: the minimum phase dynamics and the non-minimum phase dynamics. $\tilde{P}_2^+(s)$[1],

$$C_2(s) = \frac{f(s)}{\tilde{P}_2^-(s)}, \tag{2.7}$$

where$f(s) = \frac{1}{(1+\eta s)^n}$ is the low-pass filter. n is chosen in such a way that the controller becomes realizable. $\tilde{P}_2^-(s)$ is the invertible part of a process model. They η can be selected based on maximum sensitivity [1],

$$\eta = \frac{1.508 - 0.451M_s}{1.45M_s - 1.5}t_d, \tag{2.8}$$

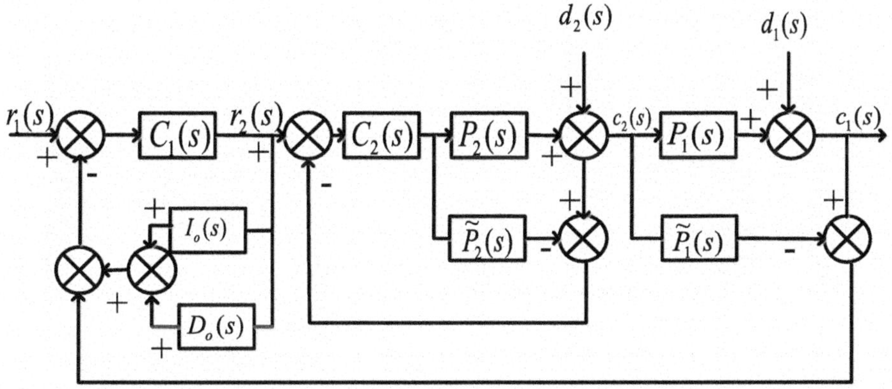

Fig. 2.2 Suggested series cascade control scheme

where t_d is the time delay term.

Outer loop controller design: Fig. 2.2, displays the resultant of the inner loop. The consequence is,

$$H(s) = \frac{C_2(s)\tilde{P}_2(s)}{1 + C_2(s)(P_2(s) - \tilde{P}_2(s))}. \tag{2.9}$$

Consequently, the plant transfer function becomes,

$$P(s) = H(s)\tilde{P}_1(s). \tag{2.10}$$

The simplified scheme is given in Fig. 2.3 [9].

$$C_1(s) = \frac{\tilde{P}^+(s)e^{-t_d s}}{\tilde{P}(s)e^{-t_d s}\{1 + \phi s^{\lambda+1} - I(s) - \tilde{P}^+(s)\} - \tilde{P}^+(s)\tilde{P}(s)(1 - e^{-t_d s})}, \tag{2.11}$$

Fractional filter parameters can be chosen based on Bode's ideal transfer function [9],

$$\lambda = \frac{\pi - PM}{\frac{\pi}{2}} - 1, \phi = \frac{1}{(\omega)^{\lambda+1}}. \tag{2.12}$$

The approach entails creating the outer loop controller for series cascade control using three time delay approximations: Pade, Taylor series, and all pole, and incorporating both an inverse response and a dead-time compensator,
Taylor series approximation [3] $e^{-t_d s} = 1 - t_d s$,
Pade approximation [4] $e^{-t_d s} = \frac{1-0.5t_d s}{1+0.5t_d s}$.

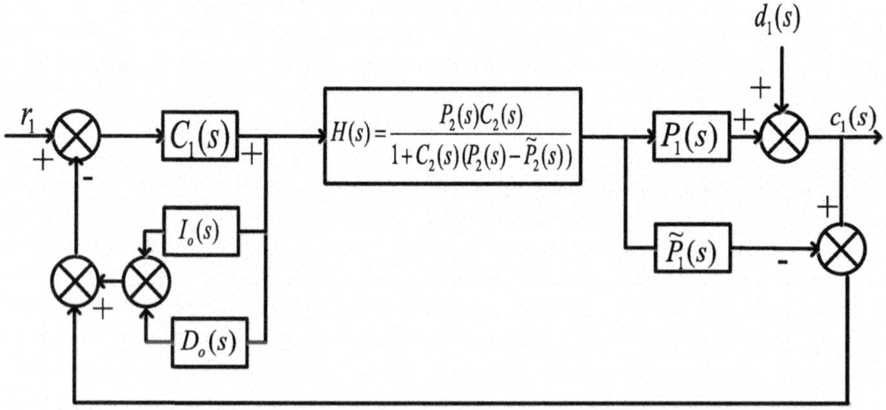

Fig. 2.3 Simplified version of the proposed scheme

All pole approximation [10] $e^{-t_d s} = \frac{1}{0.5t_d^2 s^2 + t_d s + 1}$,

In Fig. 2.3, a simplified structure of the proposed scheme is offered. The plant transfer function is given as,

$$\tilde{P}_1(s) = \frac{K_1'(1-as)}{s+b} e^{-t_{d1}s}, \tilde{P}_2(s) = \frac{K_2' e^{-t_{d2}s}}{s+c}. \tag{2.13}$$

After the utilization of Eq. 2.7, we arrived at the inner loop controller transfer function,

$$C_2(s) = \frac{(s+c)}{K_2'(1+\eta s)}. \tag{2.14}$$

Now Eq. (2.8) can be used for tuning η. After considering Eq. (2.9), the outcome of the inner loop becomes,

$$H(s) = \frac{e^{-t_{d_2}s}}{1+\eta s}. \tag{2.15}$$

From Eq. (2.10),

$$\tilde{P}(s) = \frac{K_1'(1-as)e^{-t_d s}}{(s+b)(1+\eta s)}. \tag{2.16}$$

Now Substituting the value of $\tilde{P}^+(s) = (1-as)(1-t_d s)$, $\tilde{P}(s)$, $I_o(s) = 2as\tilde{P}^-(s)$ and $D_o(s) = \frac{1-e^{-t_d s}}{e^{-t_d s}}\tilde{P}^-(s)$ in Eq. (2.11). Here dubious delay term is approximated according to Taylor series approximation, i.e., $e^{-t_d s} = 1 - t_d s$. After that, the controller transfer function becomes,

$$C_1'(s) = \frac{1}{\phi s^\lambda - a}\left(\frac{\eta s}{K_1'} + \frac{(1+b\eta)}{K_1'} + \frac{b}{K_1' s}\right). \tag{2.17}$$

Here $\frac{1}{\phi s^\lambda - a}$ demonstrate the fractional filter and indicate the PID controller.

However, the time delay term can be approximated via Pade approximation, i.e., $e^{-t_d s} = \frac{1-0.5t_d s}{1+0.5t_d s}$.

Subsequently, the transfer function of the controller is formulated as follows:

$$C_1''(s) = \frac{1+0.5t_d s}{\phi s^\lambda + 0.5as - (a+0.5t_d)}\left(\frac{(1+bs)}{K_1'} + \frac{b}{K_1'} + \frac{\eta s}{K_1'}\right). \tag{2.18}$$

The same procedure is utilized for controller design. In all pole approximation, the delay term is approximated as $e^{-t_d s} = \frac{1}{0.5t_d^2 s^2 + t_d s + 1}$.

$$C_1'''(s) = \frac{0.5t_d^2 s^2 + t_d s + 1}{\begin{array}{l}0.25at_d^4 s^4 + \phi s^\lambda (0.5t_d^2 s^2 + t_d s + 1) + s^3 (at_d^3 - 0.3t_d^4) \\ +s^2 (0.5at_d^2 - 0.5t_d^3) - s(at_d + t_d^2)\end{array}} \left(\frac{(1 + bs)}{K_1'} + \frac{b}{K_1' s} + \frac{\eta s}{K_1'} \right). \tag{2.19}$$

Significance of Dead-time and Inverse response compensator:

If the process transfer function contains only a delay term and exhibits non-minimum phase zeros, the controller transfer function takes the form of [9],

$$C_1(s) = \frac{1}{\tilde{P}^-(s)[(1 + \phi s^{\lambda+1}) - \tilde{P}(s)] - D_o(s)}.$$

The impact of the Smith predictor is depicted in Fig. 2.4.

The corresponding feedback loop in Fig. 2.5 shows the controller operating on the signal from the system lacking incorporating the Smith predictor. After integrating the Smith predictor, the modified controller structure is depicted in Fig. 2.5.

After applying the Smith predictor, the new closed-loop transfer function is:

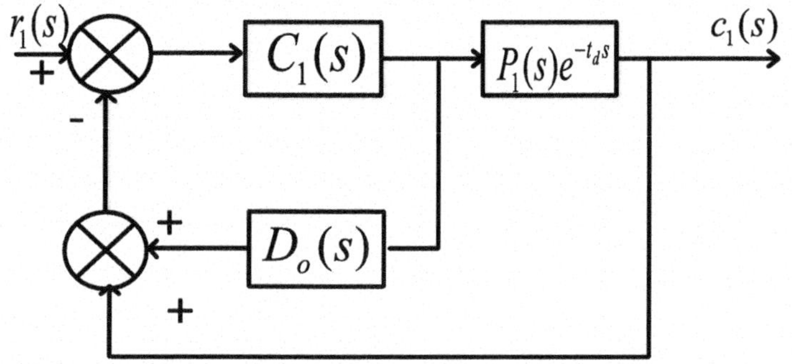

Fig. 2.4 Feedback loop with dead-time compensator

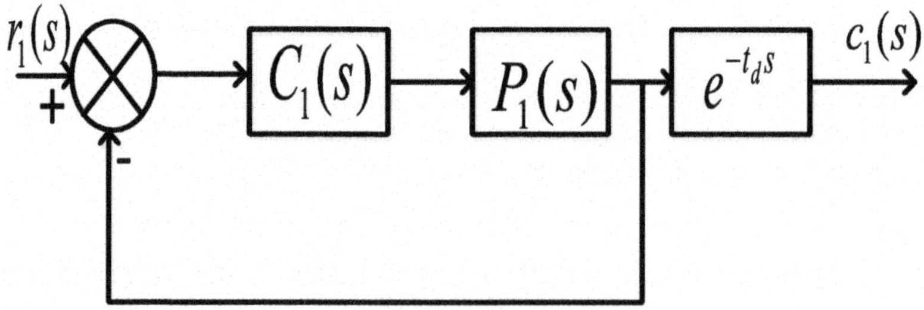

Fig. 2.5 Equivalent feedback loop with dead-time compensator

$$\frac{c_1(s)}{r_1(s)} = \left(\frac{C_1(s)P_1(s)}{1 + C_1(s)P_1(s)}\right)e^{-t_d s}.$$

After ignoring the Smith predictor term, the overall closed-loop transfer function becomes,

$$\frac{c_1(s)}{r_1(s)} = \left(\frac{C_1(s)P_1(s)e^{-t_d s}}{1 + C_1(s)P_1(s)e^{-t_d s}}\right).$$

Specific case:

Consider the SOPDT (Second Order Plus Time Delay) process transfer function,

$$P(s) = \frac{K_1' e^{-t_d s}}{(s+b)(1+\eta s)}.$$

Incorporate the Smith predictor into the controller transfer function by using the Pade approximation for the time delay and applying the Internal Model Control (IMC) technique [9].

$$C_1(s) = \frac{1 + 0.5t_d s}{(\phi s^{\lambda} - 0.5t_d)}\left(\frac{(1+bs)}{K_1'} + \frac{b}{K_1' s} + \frac{\eta s}{K_1'}\right).$$

For the same specific case (SOPDT) without the Smith predictor, the controller transfer becomes,

$$C_1(s) = \frac{(1 + 0.5t_d s)e^{-t_d s}}{\phi s^{\lambda+1} + (1 + 0.5t_d s)e^{-t_d s}}.$$

Now, to eliminate the stimulus of the time delay term, consider the process transfer function incorporating non-minimum phase (NMP) zeros, which can be articulated as [1],

$$P(s) = P'(s)(1 - as),$$

where, $P'(s) = \frac{1}{(s+b)(\eta s+1)}$
From Fig. 2.6.

$$I_o(s) = P'(s)\delta s.$$

Let us outline the variable,

$$c' = I_o(s)u(s).$$

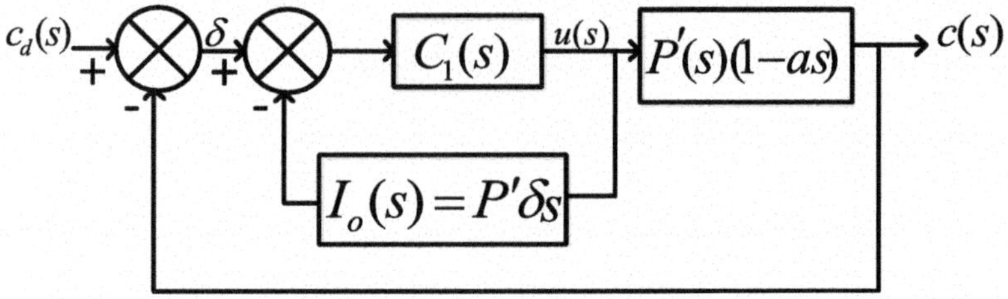

Fig. 2.6 Block diagram incorporating the inverse response compensator

Consequently, the signal reaches the controller through this minor loop, referred to as δ. It can be denoted as:

$$\delta = c_d - c(s) - c'(s). \tag{2.20}$$

Now substitute the value of $c(s)$ and $c'(s)$ into Eq. (2.20). Then it becomes,

$$\delta = c_d - [P'(s) + I_o(s)]u(s).$$

Now consider,

$$P^* = P'(s) + I_o(s).$$

or

$$P^* = P'(s)(1 - as) + P'(s)\delta s.$$

or

$$P^* = P'(s)[(1 + (\delta - a)s].$$

2.1.4 *Case Studies*

To measure the efficacy of the recommended control structure, two benchmark problems are applied.

2.1.4.1 Case Study I

Maintaining the boiler drum level is critical; if it drops too low and additional feed water is introduced, the drum level initially decreases before rising. For an in-depth understanding of the transfer function of the boiler drum level system, consult process control literature and specialized engineering resource [9].

$$\tilde{P}_1(s) = \frac{(1-0.418s)e^{-0.03s}}{s}, \tilde{P}_2(s) = \frac{0.547e^{-0.07s}}{1.06s+1}.$$

Inner loop Controller design: From Eq. (2.14),

$$C_2(s) = \frac{(1+1.06s)}{0.547(1+2.45s)},$$

where we considered the maximum sensitivity $M_s = 1.35$ for tuning of filter parameters ϕ, after utilization Eq. (2.8), we obtained $\eta = 2.45$.

Outer loop controller design: After considering Eq. (2.16),

$$\tilde{P}(s) = \frac{(1-0.418s)e^{-0.1s}}{s(1+2.45s)}. \tag{2.21}$$

After the utilization of (2.17), (2.18), and (2.19), the outer loop controller with different approximations is demonstrated in Table 2.1.

Here $PM = 49.2, \lambda = 0.6$ and $\phi = 35$. Here we have approximated the delay term for the controller design via three approximations. Figure 2.7 shows a comparative analysis of the boiler drum level system.

The Pade approximation for the time delay term leads to improvements in the outer loop controller's overshoot, undershoot, and settling time, outperforming other approximations in both the nominal case and the +10% mismatch case.

As illustrated in Table 2.2, the proposed controller exhibits reduced IAE and ISE values across both scenarios: the nominal case and the +10% mismatch case.

The suggested controller is associated with the controllers offered in [8, 9] using the same set of parameters. Figure 2.8 illustrates this comparison, highlighting significant improvements with the proposed controller.

In Fig. 2.8, the disturbances of magnitude 0.2 at t = 20 s are applied for both cases.

Table 2.1 Outer loop controller design with different approximations

Approximations	Fractional filter + PID controller
Taylor series	$C_1' = \frac{1}{35s^{0.6}-0.418}(1+2.45s).$
Pade	$C_2'' = \frac{(1+0.5s)}{35s^{0.6}+0.0209s-0.418}(1+2.45s).$
All pole	$C_3''' = \frac{(0.005s^2+0.1s+1)}{0.00001045s^4+35s^{2.6}+3.5s^{1.6}+35s^{0.6}+0.000393s^3+0.00159s^2-0.0518s}(1+2.45s).$

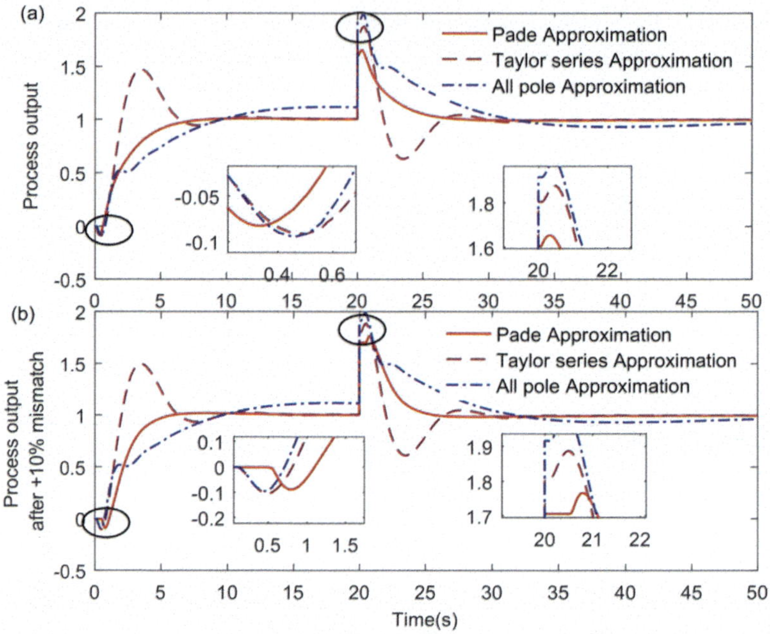

Fig. 2.7 Relative investigation with three approximations

Table 2.2 Performance indices for boiler drum level system

Controller	Nominal case		+10% mismatch case	
	IAE	ISE	IAE	ISE
Suggested controller	0.28	0.082	0.29	0.084
Ranganayakulu et al. [8]	0.40	0.16	0.43	0.18
Nagarsheth and Sharma [9]	0.68	0.46	0.86	0.74

To evaluate the stability of the proposed method, the Riemann surface is employed in conjunction with the graphical analysis provided by the FOMCON toolbox in MATLAB. The formulation of the fractional-order characteristic polynomial is detailed in [1], and Table 2.3 outlines the structure of the fractional characteristic polynomial for the proposed closed-loop system.

The precise format of the fractional characteristic polynomial for the boiler drum level system is presented in Table 2.4.

Notably, the stability analysis of the suggested closed-loop system using fractional filters can be reformulated with fractional characteristic polynomials.

The fractional characteristics polynomial has positioned the roots in $s - plane$. Here $s = \alpha + j\omega$. The mapping of $s - plane$ a plane with the equivalent aspect is available in Table 2.5. As depicted in Fig. 2.9, all roots of the polynomial in question adhere to the stability condition, $|\lambda_i = (\eta\pi/2)|$. where $\beta = 0.01$.

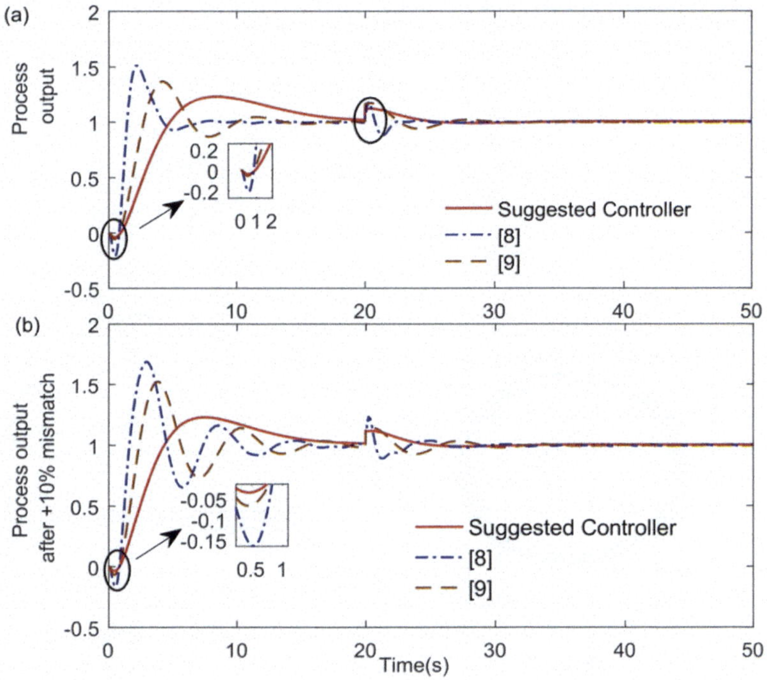

Fig. 2.8 Relative analysis for boiler drum level system

Table 2.3 Fractional characteristic polynomial in its generalized form

Approximations	Generalized fractional-order polynomial
Taylor series	$f(s) = 1 + \phi s^{\lambda+1} - \frac{t_d s}{2}$
Pade	$f(s) = 1 + \phi s^{\lambda+1} + \frac{a t_d s^2}{2} - (a + \frac{t_d}{2})s$
All pole	$f(s) = 1 + 0.25 a t_d^4 s^5 + s^4(a t_d^3 - 0.25 t_d^4)$ $s^3(0.5 a t_d^2 - t_d^3) + s^2(0.5 t_d - a t_d - 1.5 t_d^2)$ $+\phi s^{\lambda+1}(0.5 t_d^2 + t_d s + 1)$

Table 2.4 Fractional-order characteristic polynomial for boiler drum level system

Approximations	Fractional-order polynomial for boiler drum level system
Taylor series	$f(s) = 1 + 35 s^{1.6} - 0.05 s$
Pade	$f(s) = 1 + 35 s^{1.6} + 0.0209 s^2 - 0.468 s$
All pole	$f(s) = 1 + 0.175 s^{3.6} + 0.00001045 s^5 + 3.5 s^{2.6}$ $+ 35 s^{1.6} + 0.00039 s^4 + 0.00159 s^3 - 0.0518 s^2$

Table 2.5 Mapping of fractional-order characteristic polynomial

Approximations	Mapping of fractional-order polynomial
Taylor series	$f(\alpha^{1/\beta}) = 1 + 35\alpha^{1600} - 0.05\alpha^{1000}$
Pade	$f(\alpha^{1/\beta}) = 1 + 35\alpha^{1600} + 0.029\alpha^{1000} - 0.468\alpha^{1000}$
All pole	$f(\alpha^{1/\beta}) = 1 + 0.175\alpha^{3600} + 0.00001045\alpha^{5000}$ $+3.5\alpha^{2600} + 35\alpha^{1600} + 0.00039\alpha^{4000} + 0.00159\alpha^{3000}$ $-0.0518\alpha^{2000}$

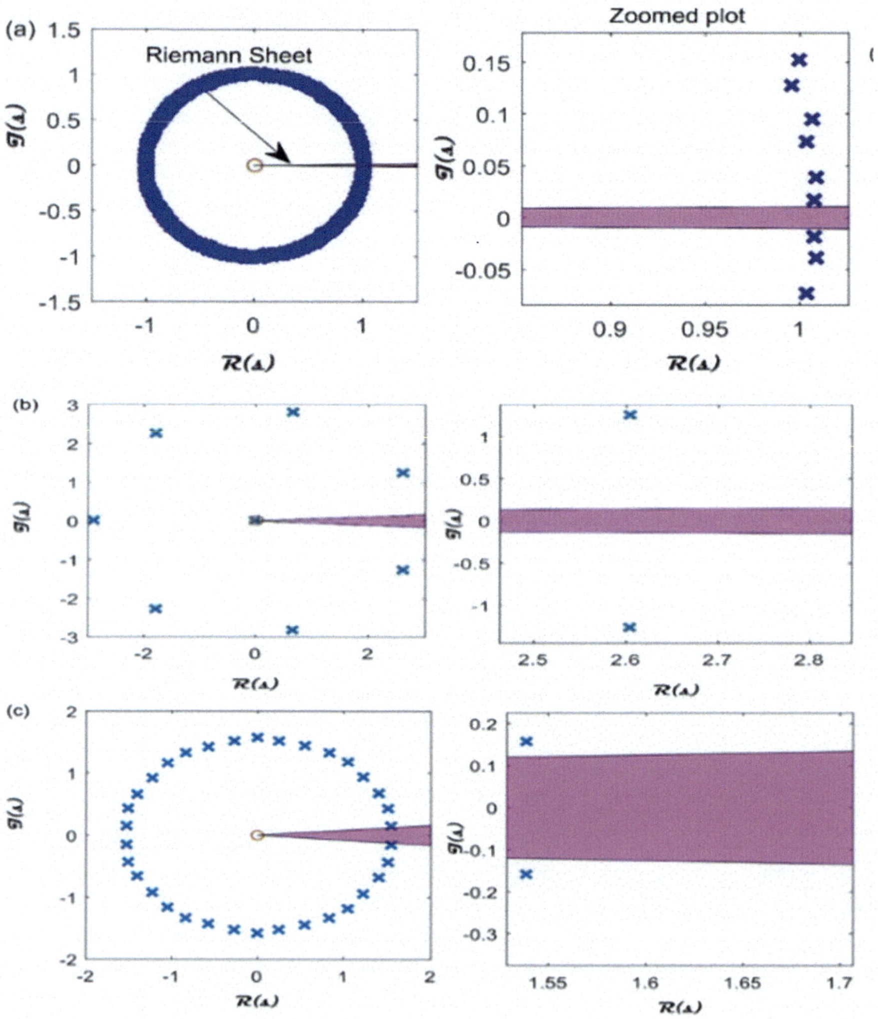

Fig. 2.9 Stability analysis of boiler drum level system

A quasi-characteristic polynomial of natural degree has stable roots if they are positioned within the Riemann surface, reflecting system stability. Conversely, instability is inferred if these roots lie outside the Riemann surface [9]. Figure 2.9 provides a graphical depiction illustrating the stability assessment of the fractional characteristic polynomial.

Furthermore, sensitivity analysis is a valuable tool for quantifying the robustness of a controller against variations in process parameters and input disturbances.

From Eq. (2.21) and the controller transfer function of Pade approximation from Table 2.1. The sensitivity function is formulated as,

$$S(s) = \frac{1}{1 + P(s)C(S)}. \tag{2.22}$$

Figure 2.10 demonstrates the graphical presentation of absolute sensitivity.

Remark 2.1 To evaluate the worth of the recommended technique, we equated it with two existing controllers as detailed in [8, 9], using identical system parameters. The comparison focused on metrics such as overshoot, undershoot, settling time, and robustness.

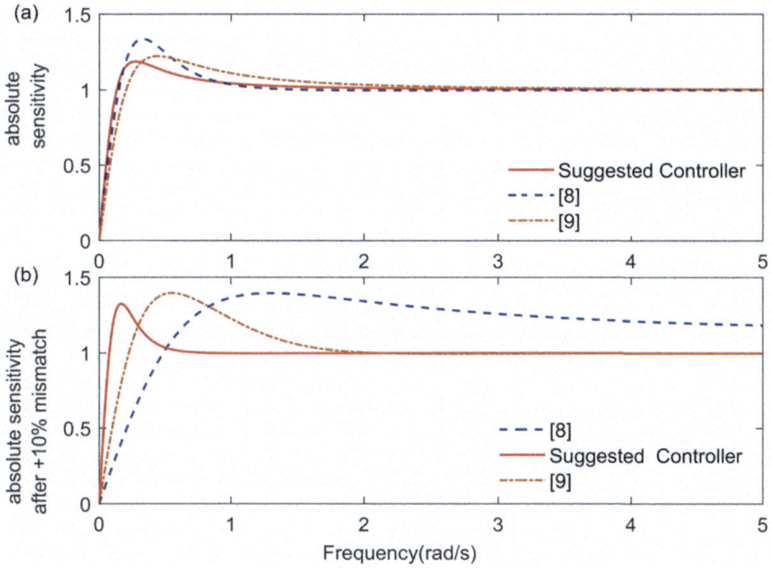

Fig. 2.10 Robustness of the boiler drum level system

2.1.4.2 Case Study II

Distillation columns typically exhibit slow dynamics and significant dead times due to factors such as composition and temperature variations. The transfer function describing the distillation column is detailed in [9].

$$\tilde{P}_1(s) = \frac{(1 - 0.2s)e^{-0.1s}}{s + 1}, \tilde{P}_2(s) = \frac{e^{-0.1s}}{s + 1}.$$

Inner loop controller design: After utilization Eq. (2.7),

$$C_2(s) = \frac{(s + 1)}{(1 + 0.3s)}.$$

Outer loop controller design: From Eq. (2.16),

$$\tilde{P}(s) = \frac{(1 - 0.2s)e^{-0.2s}}{(1 + s)(s + 3)} \tag{2.23}$$

Here Eqs. (2.17), (2.18), and (2.19) are used to design the outer loop controller. It is demonstrated in Table 2.6. Figure 2.11 presents a comparative analysis of the distillation column, highlighting that employing the Pade approximation results in the outer loop controller exhibiting reduced overshoot, undershoot, and settling time compared to other approximations.

The results, as revealed in Fig. 2.12, associate the recommended way with existing controllers from [8, 9] and demonstrate notable performance improvements in both the nominal and +10% mismatch scenarios.

In Table 2.8, the fractional characteristic polynomial pertinent to the distillation column is outlined in detail. Graphical assessment is conducted in Fig. 2.13 using the FOMCON toolbox in MATLAB, as described by [11, 12]. The depiction in Fig. 2.13 shows all roots outside the Riemann sheet, reflecting that the system is stable.

Table 2.6 Outer loop controller design for the distillation column

Approximations	Fractional filter + PID controller
Taylor series	$C_1' = \frac{1}{0.70s^{1.45} - 0.05s}(1 + s)$
Pade	$C_1''(s) = \frac{(1+0.1s)}{0.70s^{0.45}+0.20s-0.1}(1 + s)$
All pole	$C_1'''(s) = \frac{(0.005s^2+0.1s+1)}{0.00008s^4 + 0.70s^{0.45}(0.02s^2 + 0.2s + 1) + 0.00112s^3 - 0.00518s}(1 + s)$

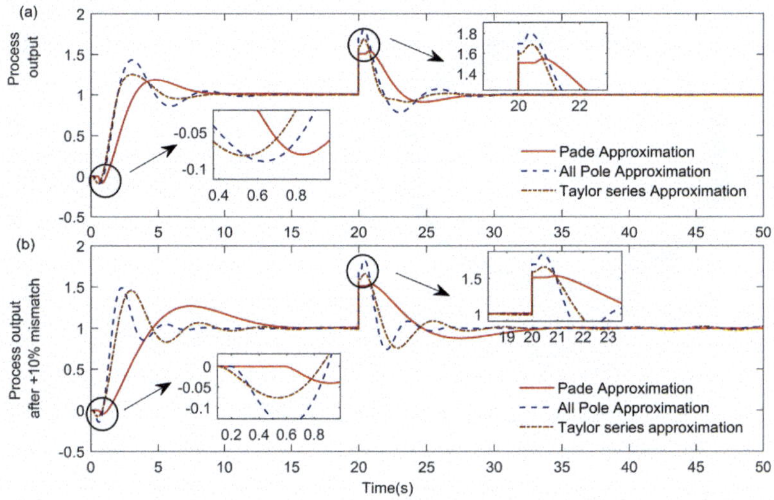

Fig. 2.11 Analysis of series cascade control for the distillation column

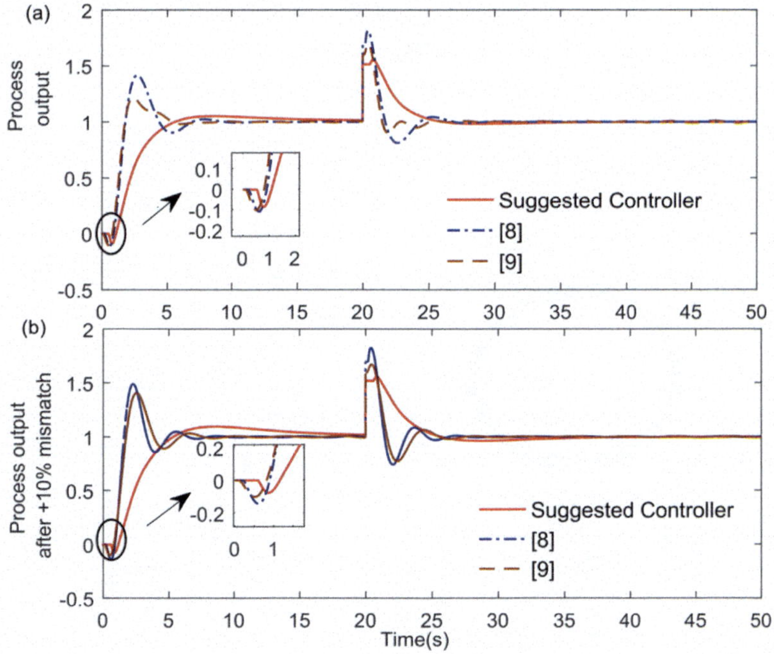

Fig. 2.12 Comparative analysis for the distillation column

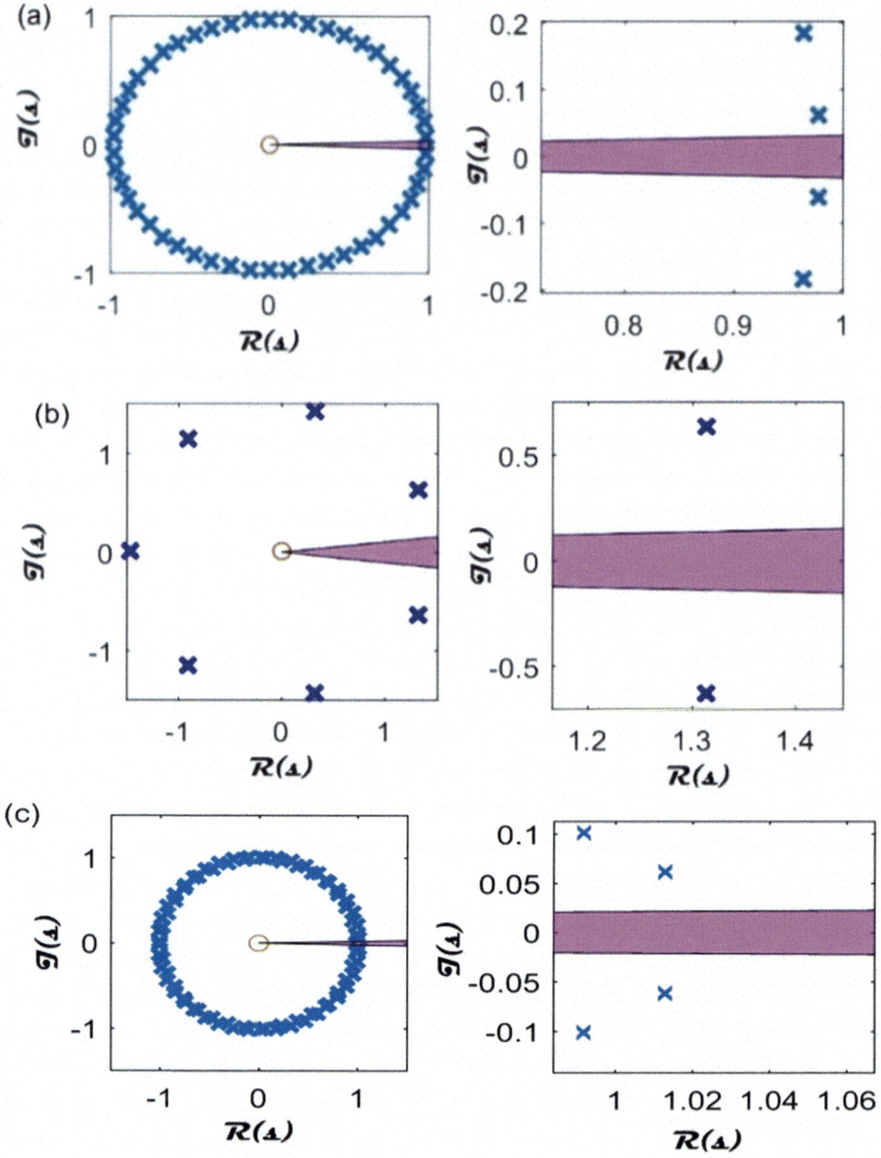

Fig. 2.13 Stability analysis of distillation column

The usefulness of the offered method is confirmed via sensitivity examination (Eq. 2.22). This analysis incorporates Eq. 2.23 and the transfer function of the controllers, detailed in Table 2.5. Figure 2.14 provides a graphical representation illustrating absolute sensitivity. Here Table 2.7 demonstrates performance indices (Table 2.8).

Remark 2.2 When compared with two existing controllers [8, 9] under identical parameters, the proposed controller shows that the existing controllers, lacking an inverse response compensator, used a setpoint filter to limit overshoot.

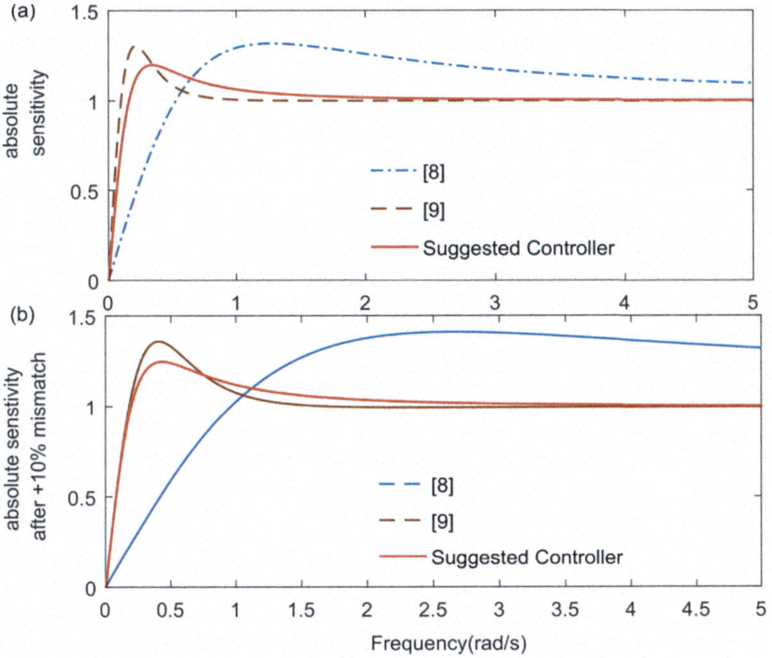

Fig. 2.14 Robustness of the controller for the distillation column

Table 2.7 Fractional-order characteristic polynomial for the distillation column

Approximations	Fractional-order polynomial for distillation column
Taylor series	$f(s) = 1 + 0.70s^{1.45} - 0.05s$
Pade	$f(s) = 1 + 0.70s^{1.45} + 0.20s^2 - 0.1s$
All pole	$f(s) = 1 + 0.000084s^5 + 0.70s^{1.45}(0.02s^2 + 0.2s + 1)$ $0.0012s^4 - 0.058s^2$

Table 2.8 Performance indices for the distillation column

Controller	Nominal case		+10% Perturbed case	
	IAE	ISE	IAE	ISE
Suggested controller	0.36	0.12	0.40	0.16
Ranganayakulu et al. [8]	0.45	0.20	0.53	0.28
Nagarsheth and Sharma [9]	0.84	0.71	1.06	1.05

2.2 Decomposed Series Cascade Control Scheme

As depicted in Fig. 2.15, the control scheme employs a higher-order fractional filter, an inverse response compensator, and a dead-time compensator, all integrated with the decomposed outer loop.

$$\begin{aligned}&c_1(s) = P_2(s)C_2(s)P_1(s)\{C_1(s)I_o(s)\}r_1(s) + (1 - P_2(s)C_2(s))\\&(1 - P_2(s)C_2(s)P_1(s)\{C_1(s)I_o(s)d_2+\\&((1 - P_2(s)C_2(s)P_1(s)C_1(s)I_o(s))d_1.\end{aligned}$$

For IMC, the inner loop controller can be defined as [1].

$$C_2(s) = \frac{f(s)}{\tilde{P}_2^-(s)}$$

Here $f(s) = \frac{1-\frac{t_d s}{2}+\frac{t_d^2 s^2}{12}}{1+\frac{t_d s}{2}+\frac{t_d^2 s^2}{12}}$ is a low-pass filter that $\tilde{P}_2^-(s)$ interprets the minimum phase part of the process model.

From Fig. 2.15,

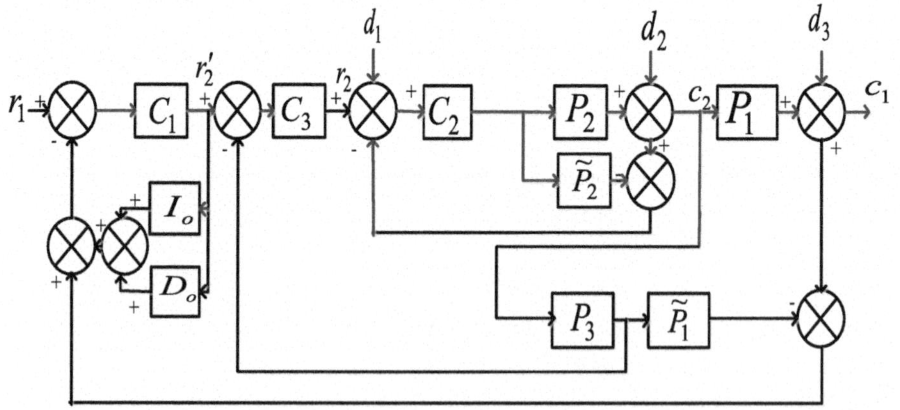

Fig. 2.15 Decomposed cascade scheme for decomposed integrating process

$$\frac{c_2(s)}{r_2(s)} = H(s) = P_2(s)C_2(s) \tag{2.24}$$

Here Fig. 2.15 represents that the outer loop is decomposed into two parts, i.e., $P_1(s)$ and $P_3(s)$.

$$P_3'(s) = H_{r_2}(s)P_3(s) \tag{2.25}$$

We can now examine the results for the decomposed portion of the outer loop, focusing on its output characteristics,

$$C_3(s) = \frac{f_3(s)}{\tilde{P}_3^-(s)} \tag{2.26}$$

The expression for the inner loop controller in the IMC context is given by [1]

$$\frac{c_2(s)}{r_2'(s)} = H_{r_3}(s) = C_3(s)P_3' \tag{2.27}$$

Afterward, the overall plant transfer function becomes (From Fig. 2.16)

$$P(s) = H_{r_3}(s)\frac{P_1(s)}{P_3(s)} \tag{2.28}$$

Now utilize the generalized controller transfer function [9]

$$C(s) = \frac{\tilde{P}^+(s)e^{-Ls}}{\tilde{P}(s)e^{-Ls}\left[f(s)^{-1}\tilde{P}(s) - \tilde{P}^+(s)\right] - 2bs - \tilde{P}(s)\tilde{P}^+(s)(1 - e^{-Ls})} \tag{2.29}$$

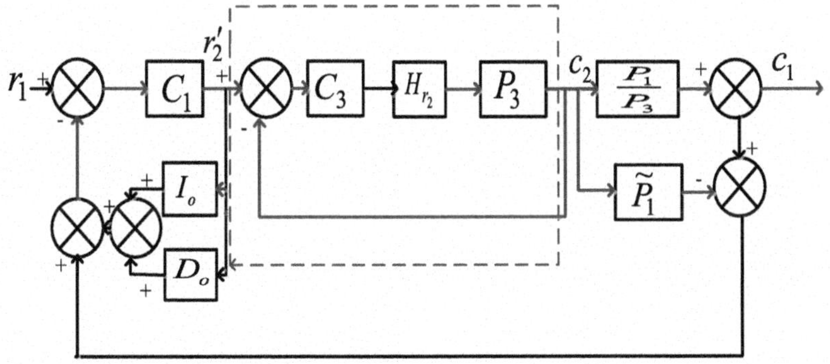

Fig. 2.16 Simplified version of decomposed scheme

Sketch of controllers for cascade loop: First order plus dead time FOPTD system in the inner loop and Double Integer order plus time delay (IOPDT) system in the outer loop.

Sketch of C_2 with FOPDT model:

$$P_2(s) = \frac{K_2' e^{-t_{d_2} s}}{(1 + T_2 s)}$$

Here K_2, t_{d2} and T_2 shows process gain, delay time, and time constant of the inner loop.

The inner loop IMC filter [8]

$$f(s) = \frac{1}{(\lambda_2 s + 1)^n}$$

where λ_2 is the tuning parameter r chosen in such a way to make the controller proper. According to the IMC notion, the process model is deteriorating into two parts minimum phase part $\tilde{P}_2^-(s) = \frac{K_2'}{1+T_2 s}$ and the non-minimum phase part $e^{-t_{d_2} s}$.

Using Eq. (2.7), we derive the inner loop IMC controller as described below,

$$C_2(s) = \frac{1 + T_2 s}{K_2(\lambda_2 s + 1)} \tag{2.30}$$

The inner loop output, following the application of Eq. (2.24), is

$$H_{r_2}(s) = \frac{e^{-t_{d2} s}}{\lambda_2 s + 1} \tag{2.31}$$

The outer loop controllers are decomposed into two parts, i.e., [12]

$$P_1(s) = \frac{K_1'(1 - as)}{s} e^{-t_{d1} s} \quad \text{and} \quad P_3(s) = \frac{K_3'}{s}$$

Design of C_3 with IO (Integer-order) model:

$$P_3(s) = \frac{K_3}{s}$$

From Eq. (2.25),

$$P_3'(s) = \frac{K_3' e^{-t_{d2} s}}{s(\lambda_2 s + 1)} \tag{2.32}$$

From Eq. (2.26),

$$C_3(s) = \frac{s(\lambda_2 s + 1)}{K_3(\lambda_3 s + 1)} \tag{2.33}$$

Here λ_3 is a tuning parameter that can be tuned by Eq. (2.27). The output of the first decomposed outer loop becomes,

$$H_{r_3}(s) = \frac{e^{-t_{d2}s}}{\lambda_3 s + 1} \tag{2.34}$$

Now the output of the integer-order model with controller $H_{r_3}(s)$ acts as an input to the non-minimum phase integer order plus delay time process model

$$P_1(s) = \frac{K_1'(1 - as)e^{-t_{d1}s}}{s}$$

After utilization Eq. (2.28),

$$P(s) = \frac{K_1'(1 - as)e^{-t_d s}}{s(\lambda_3 s + 1)} \tag{2.35}$$

Here $t_d = t_{d_1} + t_{d_2}$.

From Eq. (2.29),

$$C(s) = \frac{s(\lambda_3 s + 1)\left(1 + \frac{t_d s}{2} + \frac{t_d^2 s^2}{12}\right)}{\phi^2 s^{2\lambda+2} + 2\phi s^{\lambda+1} + \frac{a t_d^2 \tau s^4}{12} + s^3\left(\frac{a t_d \tau}{2} + \frac{a t_d^2}{12} - \frac{t_d^2 \tau}{12}\right) + s^2\left(\frac{a t_d}{2} - \frac{t_d^2}{12} - a\tau - \frac{a\tau}{2}\right) - s\left(a + \frac{t_d}{2} + \tau\right)} \tag{2.36}$$

Consequently, when equation (2.30) is split into the fractional filter and PID controller elements, it takes the form of,

$$C(s) = \frac{\lambda_3 s^2 + s}{\phi^2 s^{2\lambda+1} + 2\phi s^{\lambda} + \frac{a t_d^2 \phi s^3}{12} + s^2\left(\frac{a t_d \tau}{2} + \frac{a t_d^2}{12} - \frac{t_d^2 \tau}{12}\right) + s\left(\frac{a t_d}{2} - \frac{t_d^2}{12} - a\tau - \frac{t_d \phi}{2}\right) - \left(a + \frac{t_d}{2} + \tau\right)}\left(\frac{t_d}{2}\left(1 + \frac{1}{t_d^2 s} + \frac{t_d s}{6}\right)\right) \tag{2.37}$$

τ is a tuning parameter that can be tuned by [9]

$$\varphi = \lambda\left[1 - \sqrt{(1 - \frac{\phi}{\lambda})^3 e^{-\frac{t_d}{\lambda}}}\right].$$

Design of controllers for cascade loop: First order plus dead time FOPTD system in the inner loop and Integer order plus time delay (IOPDT) system in the outer loop.

Design of $C_2(s)$:

The same technique as used previously is applied in the design of the inner loop controller.

$$C_2(s) = \frac{1 + T_2 s}{K_2'(\lambda_2 s + 1)}$$

From Eq. (2.24) the inner loop becomes

$$H_{r_2}(s) = \frac{e^{-t_d s}}{\lambda_2 s + 1}$$

Design of $C_3(s)$: Since the integer-order controller receives the output from the inner loop transfer function as its input, the plant transfer function, after applying Eq. (2.25), becomes:

$$P_3'(s) = \frac{K_3' e^{-t_{d2} s}}{(T_3 s + 1)(\lambda_2 s + 1)} \tag{2.38}$$

From Eq. (2.26),

$$C_3(s) = \frac{s(\lambda_2 s + 1)(T_3 s + 1)}{K_3'(\lambda_3 s + 1)} \tag{2.39}$$

From Eq. (2.27)

$$H_{r_3}(s) = \frac{e^{-t_{d2} s}}{\lambda_3 s + 1}$$

Now the output of the integer-order model with controller $H_{r_3}(s)$ acts as an input to the non-minimum phase integer order plus delay time process model

$$P_1(s) = \frac{K_1'(1 - as)e^{-t_{d1} s}}{s(T_1 s + 1)}$$

From Eq. (2.28),

$$P(s) = \frac{K_1'(1 - as)e^{-t_d s}}{s(\lambda_3 s + 1)(T_1 s + 1)} \tag{2.40}$$

Therefore, the controller design in the IMC framework was discussed previously. When generalized Eq. (2.29) is used for controller design, it results in,

$$C(s) = \frac{\lambda_3 s^3 + s^2(T_1 + \lambda_3) + s}{\phi^2 s^{2\lambda+1} + 2\phi s\lambda + \frac{a t_d^2 \tau s^3}{12} + s^2\left(\frac{a t_d \tau}{2} + \frac{a t_d^2}{12} - \frac{t_d^2 \tau}{12}\right) + s\left(\frac{a t_d}{2} - \frac{t_d^2}{12} - a\tau - \frac{t_d \phi}{2}\right) - \left(a + \frac{t_d}{2} + \tau\right)} \tag{2.41}$$

2.2.1 *Case Studies*

This section confines the evaluation of the proposed scheme to processes that demonstrate both inverse response and dead time characteristics.

2.2.1.1 Case Study I

The transfer function for a double integrating process, incorporating inverse response and dead time, is given by [11]

$$P_1(s) = \frac{(1 - 0.7s)e^{-0.1}}{s^2}, \; P_2(s) = \frac{e^{-0.1}}{s + 1}.$$

From Eq. (2.30),

$$C_2(s) = \frac{1 + s}{0.4s + 1}$$

Here $\lambda_2 = 0.4$ which can be chosen on maximum sensitivity M_s.

From Eq. (2.32),

$$P_3'(s) = \frac{e^{-0.1s}}{s(0.4s + 1)}$$

From Eq. (2.33),

$$C_3(s) = \frac{s(0.4s + 1)}{0.2s + 1}$$

The first decomposed component's output acts as the input for the remaining outer loop controller. From Eq. (2.35),

$$P(s) = \frac{(1 - 0.7s)e^{-0.2s}}{s(s + 1)} \tag{2.42}$$

The outer loop controller settings in the fractional IMC framework are derived from the guidelines provided by Eq. (2.36).

$$C(s) = \frac{0.3s^2 + s}{s^{1.90} + 2s^{0.45} + 0.000067s^3 + 0.0023s^2 - 0.013s - 0.4}\left(0.1 + \frac{2}{s} + 0.3s\right) \tag{2.43}$$

The expression $\frac{0.3s^2+s}{s^{1.90} + 2s^{0.45} + 0.000067s^3 + 0.0023s^2 - 0.013s - 0.4}$ depict the fractional IMC filter and $\left(0.1 + \frac{2}{s} + 0.3s\right)$ elucidate the PID controller. Here, $\lambda_3 = 0.5\,\theta = 0.1$, $\lambda = 1$,and $\eta = 0.45$ with $PM = 45$. A comparative analysis of the closed-loop response, using the recommended approach, is presented in Fig. 2.17 (Table 2.9).

The structure of the fractional characteristic polynomial pertinent to the boiler drum level's closed-loop system is outlined as [9].

$$f(s) = 1 + s^{2.90} + 2s^{1.45} + 0.000067s^4 + 0.0023s^3 - 0.013s^2 - 0.4s\,.$$

The fractional-order characteristic polynomial roots lie in mapping a plane linked with an equivalent characteristic polynomial. Choose $\alpha = s^{\gamma}$, where $s^{0.001} = \alpha$.

$$f(s) = 1 + \alpha^{2900} + 2\alpha^{1450} + 0.000067\alpha^{4000} + 0.0023\alpha^{3000} - 0.013\alpha^{2000} - 0.4\alpha^{1000}.$$

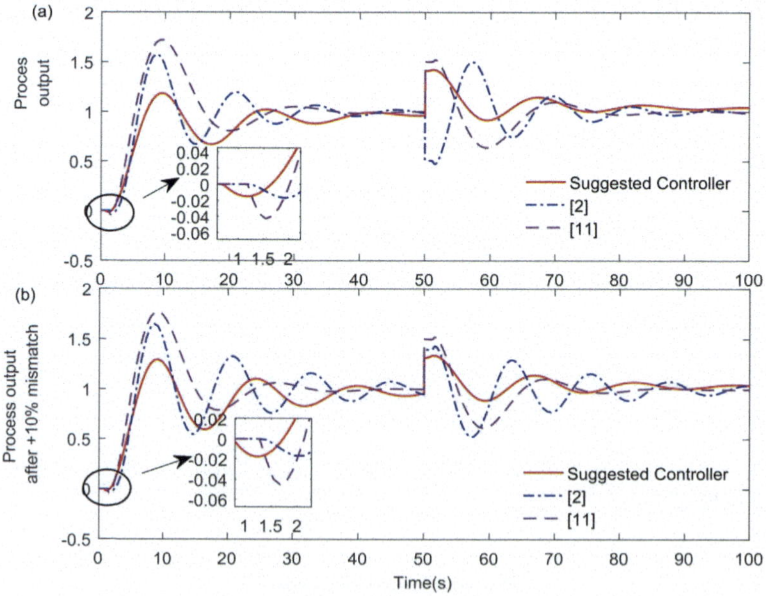

Fig. 2.17 Relative investigation of double integrating process

Table 2.9 Transient response and performance indices for double integrating process

Method	Nominal case				+10% mismatch case			
	O.S.	Rise time	IAE	ISE	O.S.	Rise time	IAE	ISE
Suggested controller	1.18	9.4	0.17	0.00	1.29	4.6	0.04	0.002
Begum et al. [2]	1.53	2.45	0.06	0.02	1.6	3.1	0.13	0.017
Kaya [11]	1.68	2.32	3.4	11.9	1.7	2.2	3.6	13.5

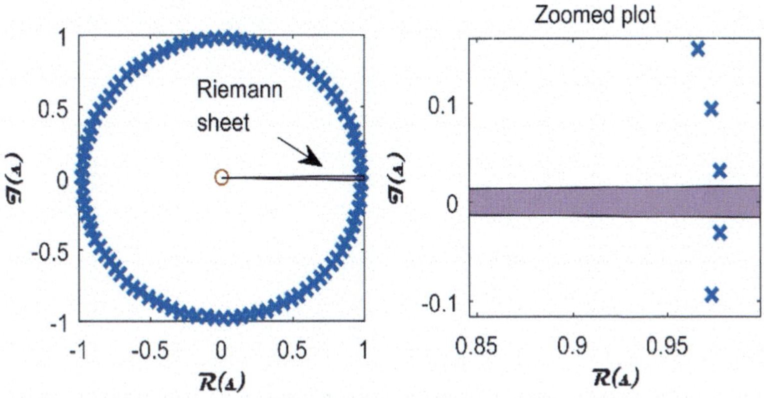

Fig. 2.18 Stability analysis of double integrating process

According to the natural degree quasi-characteristic polynomial, instability is suggested by roots within the Riemann sheet, whereas stability is indicated by roots located outside. Figure 2.18 illustrates that all roots are positioned outside the Riemann sheet. It is also important to recognize that variations in process dynamics and disturbances could influence the stability assessments of the feedback system.

The universal sensitivity function can be transcribed as [9]

$$S(s) = \frac{1}{1 + \tilde{P}(s)C(s)} \tag{2.44}$$

To advance, input the values from Eqs. (2.20) and (2.21) into Eq. (2.22). The robustness analysis illustrated in Fig. 2.19 reveals that the proposed controllers experience a smaller maximum sensitivity increase in both the nominal case and when subjected to a 10% model mismatch.

Remark 2.3 In evaluating the proposed controller, a comparison is made with the strategies outlined in [2, 11]. The approach in [2] involves a setpoint filter for managing overshoot, and [11] uses a lower-order Pade approximation. Conversely, this manuscript applies both a higher-order fractional IMC filter and a higher-order

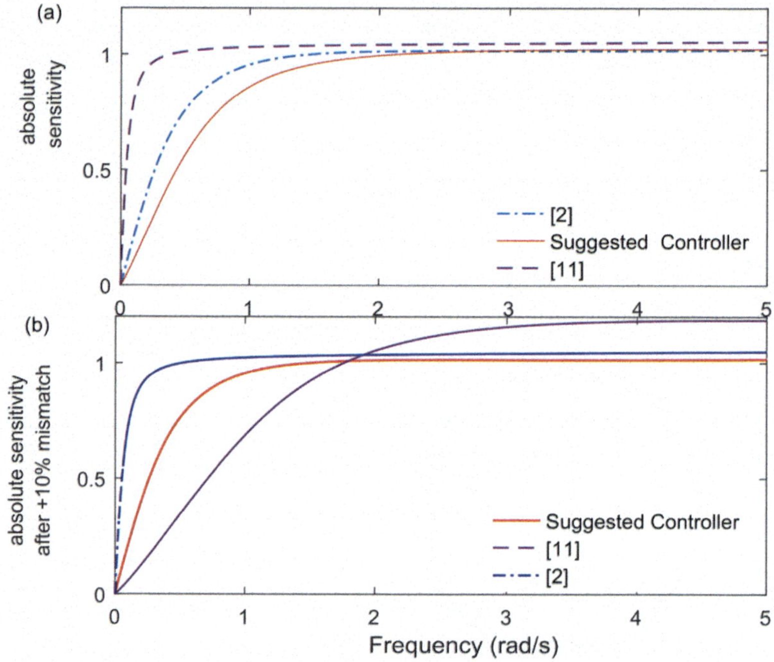

Fig. 2.19 Robustness analysis of decomposed controller for the double integrating process

Pade approximation. This approach avoids the need for a setpoint filter and reduces the magnitude jump under nominal conditions and a +10% model mismatch with disturbances.

2.2.1.2 Case Study II

Consider a higher-order process transfer function incorporating inverse response along with dead- time [11].

$$P_1(s) = \frac{(1 - 0.5s)e^{-0.4s}}{s(0.5s + 1)(0.4s + 1)}, \; P_2(s) = \frac{0.5e^{-034s}}{(0.1s + 1)}.$$

After applying Eq. (2.7), the transfer function of the inner loop controller is,

$$C_2(s) = \frac{0.1s + 1}{0.5(s + 1)}$$

From Eq. (2.31), the output of the inner loop becomes

$$H_{r_2}(s) = \frac{e^{-0.3s}}{s+1}$$

From Eq. (2.38), the plant transfer function becomes

$$P'_3(s) = \frac{e^{-0.3s}}{(s+1)(0.4s+1)}$$

From Eq. (2.39)

$$C_3(s) = \frac{(s+1)(0.4s+1)}{(0.2s+1)}$$

Now, the output from the first decomposed section of the outer loop is

$$H_{r_3}(s) = \frac{e^{-0.3s}}{0.2s+1}$$

From Eq. (2.40),

$$P(s) = \frac{(1-0.5s)e^{-0.7s}}{s(0.5s+1)(0.2s+1)} \tag{2.45}$$

As previously discussed in Sect. 2.2, after utilization of Eq. (2.41) structure of the controller transfer function becomes,

$$C(s) = \frac{0.20s^3 + 0.9s^2 + s}{0.25s^{1.90} + s^{0.45} + 0.0020s^3 + 0.05s^2 + 0.045s - 0.95}\left(0.35 + \frac{0.39}{s} + 0.038s\right) \tag{2.46}$$

The term represents the fractional IMC filter and $\left(0.35 + \frac{0.39}{s} + 0.038s\right)$ interprets the PID controller. Here, $\lambda_3 = 0.5\ \theta = 0.1$, $\lambda = 1$,and $\eta = 0.45$ with $PM = 45$. The performance indices are revealed in Table 2.10. Figure 2.20 shows comparative study of the closed-loop response formed from the recommended approach. Figure 2.20 provides a comparative analysis of the closed-loop response using the recommended approach. In Fig. 2.20a, a fractional filter-based IMC controller is combined with an inverse response and dead-time compensator for the second decomposed segment of the outer loop.

Figure 2.20b indicates that, following a +10% model mismatch, the proposed control scheme delivers convincing results in terms of both overshoot and undershoot. Table 2.10 shows transient response and performance indices.

Additionally, the precise arrangement of the fractional characteristic polynomial concerning the closed-loop of the higher-order integrating process is depicted.

$$f(s) = 1 + s^{2.90} + 2s^{1.45} + 0.000067s^4 + 0.0023s^3 - .013s^2 - 0.4s\,.$$

Table 2.10 Transient response and performance indices for higher-order integrating process

Method	Nominal case				+10% mismatch case			
	O.S.	Rise time	IAE	ISE	O.S.	Rise time	IAE	ISE
Suggested controller	1.17	30	0.012	0.0001	1.32	32.8	0.013	0.0001
Begum et al. [2]	1.2	14.66	0.08	0.0072	1.43	18.5	0.95	0.91
Kaya [11]	1.46	5.8	1.4	2	1.5	7.8	1.5	2.4

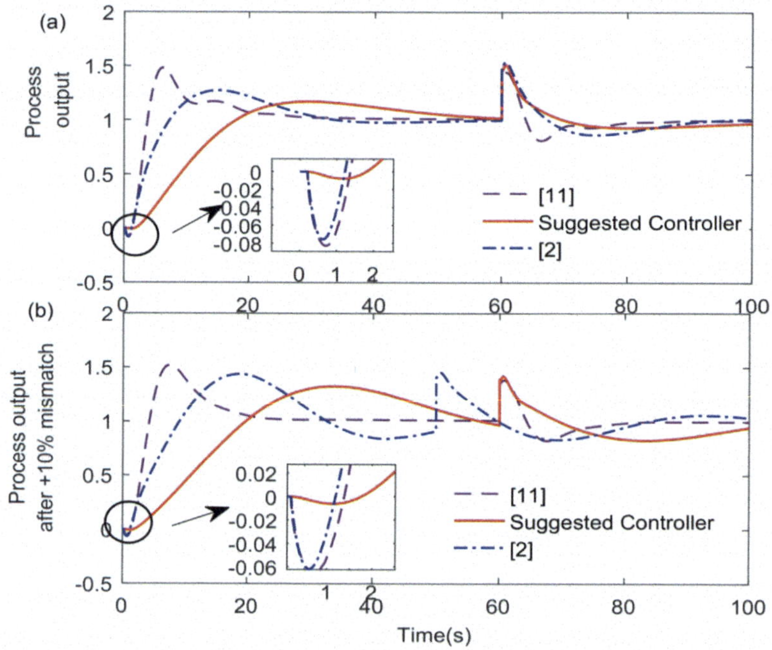

Fig. 2.20 Comparative study of higher-order integrating process

The stability of the system is confirmed by Fig. 2.21, where all roots are found outside the Riemann sheet.

Figure 2.22 provides an evaluation of the proposed controller's robustness, utilizing Eqs. (2.22), (2.23), and (2.24) for the analysis. The controllers are shown to have a lower peak sensitivity in nominal scenarios as well as with a 10% model mismatch.

Remark 2.4 Figure 2.22 illustrates the effectiveness of the proposed controller by comparing it with methods discussed in [2, 11]. While [11] uses a setpoint filter to control overshoot and [2] relies on a lower-order Pade approximation, this manuscript adopts a higher-order fractional IMC filter and a higher-order Pade approximation,

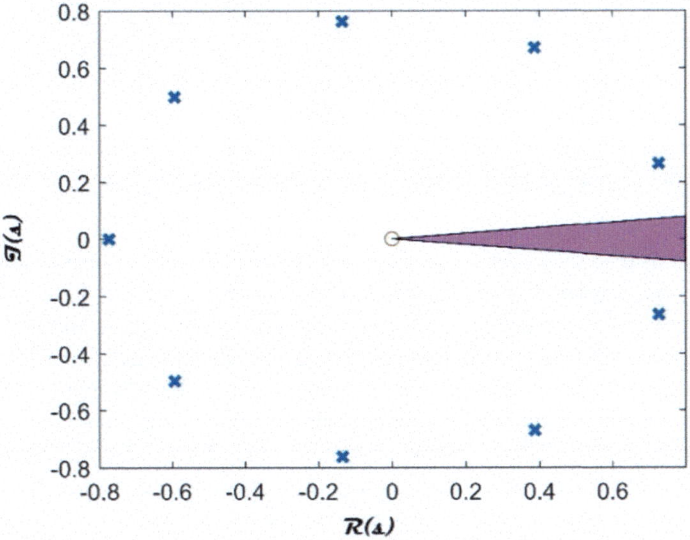

Fig. 2.21 Stability analysis of higher-order integrating process

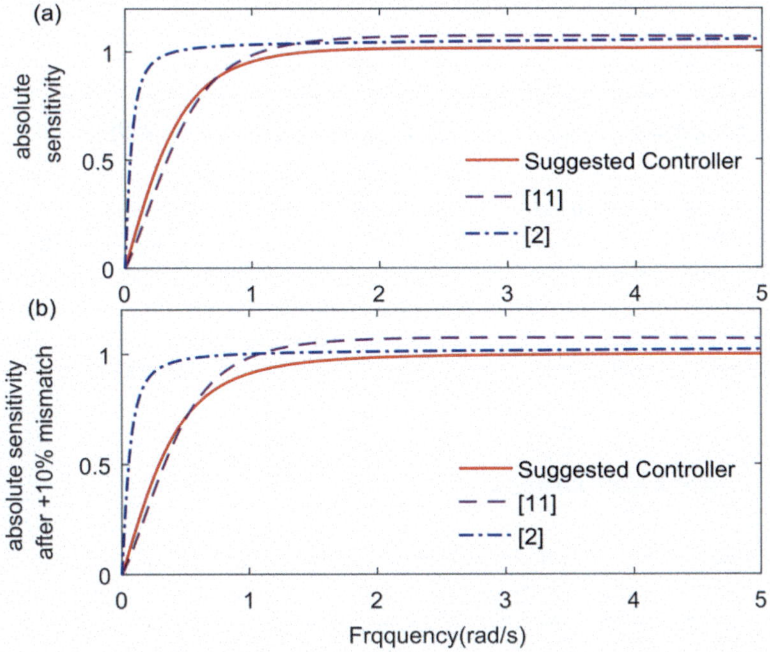

Fig. 2.22 Robustness analysis of decomposed controller for higher-order integrating process

omitting the setpoint filter and achieving a reduced magnitude jump in nominal conditions as well as with a $+10\%$ model mismatch and disturbances.

2.3 Improved Control Structure with Noise Filter

A limitation of feedback systems is that measurement noise can lead to undesirable control actions, potentially causing actuator wear or even system failure. This section aims to improve control arrangements by integrating a noise filter into the feedback path of the proposed control structure (see Fig. 2.15). Measurement noise often enters the system through various paths, typically assumed to be additive at the process output [13]. Effective filtering is crucial, particularly for controllers with derivative action, to mitigate the effects of measurement noise on the control signal within acceptable limits. The design of the measurement noise filter is intended to reduce or moderate control responses induced by measurement noise while maintaining overall system performance and robustness. The evaluation of control signal deviations and filtered outputs can be performed using the transfer functions of the process and controller, considering factors like the spectral density of the measurement noise [14].

2.3.1 Modified Series Cascade Control with Noise Filter

To mitigate the effects of noise, a noise filter has been incorporated. Section 2.1.3 offers a comprehensive explanation of the proposed scheme, including this noise filter. This enhanced version of the original system is illustrated in Fig. 2.23. Although Sect. 2.1 of the literature review notes that outer loop controllers often use fractional filters and Smith predictors, it highlights that noise considerations are typically omitted.

2.3.2 Noise Filter Design

Adding a noise filter can affect both the effectiveness and resilience of the control system. Integrating the filter dynamics with the process dynamics might necessitate returning the controller to maintain a balance between performance and robustness. On the other hand, adjustments to the controller's settings might also require corresponding changes to the noise filter to ensure continued optimal performance. This study proposes a revised approach that integrates considerations of performance, robustness, and noise attenuation [14] suggests an iterative process to resolve similar issues. Where the filter time constant T_m is selected as,

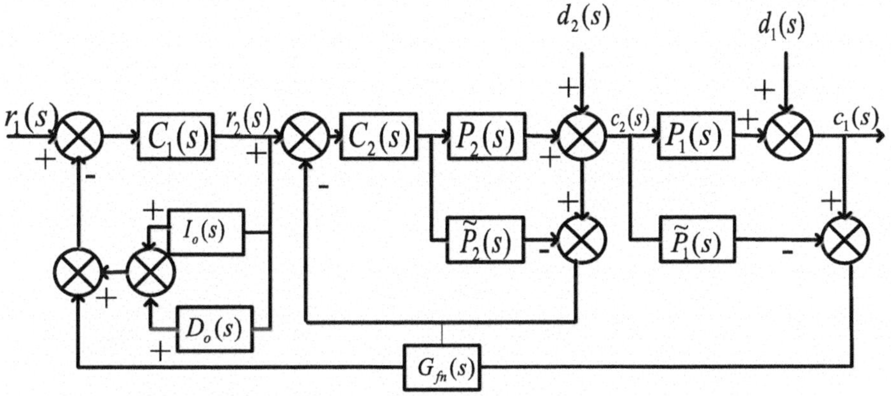

Fig. 2.23 Proposed control scheme with noise filter

$$T_m = \frac{\psi}{\omega}. \tag{2.24}$$

Controllers featuring a range of filter time constants will be analyzed to explore their performance variations [11].

$$G_{fn}(s) = \frac{1}{T_m^2 s^2 + T_m s + 1}. \tag{2.25}$$

2.3.3 *Case Studies*

To evaluate the suggested control scheme two benchmark problems are used.

2.3.3.1 Case Study I

The process transfer function of the boiler drum level system is given [8]

$$\tilde{P}_1(s) = \frac{(1 - 0.418s)e^{-0.07s}}{s}, \tilde{P}_2(s) = \frac{0.547e^{-0.03s}}{1.06s + 1}.$$

After utilization Eqs. (2.24) and (2.25), the noise filter becomes,

$$G_{f_1}(s) = \frac{1}{0.0069s^2 + 0.0836s + 1}.$$

Here $\omega = 0.26$ is the loop gain frequency. Here $T_m = 0.0836$ with $\psi = 0.0604$

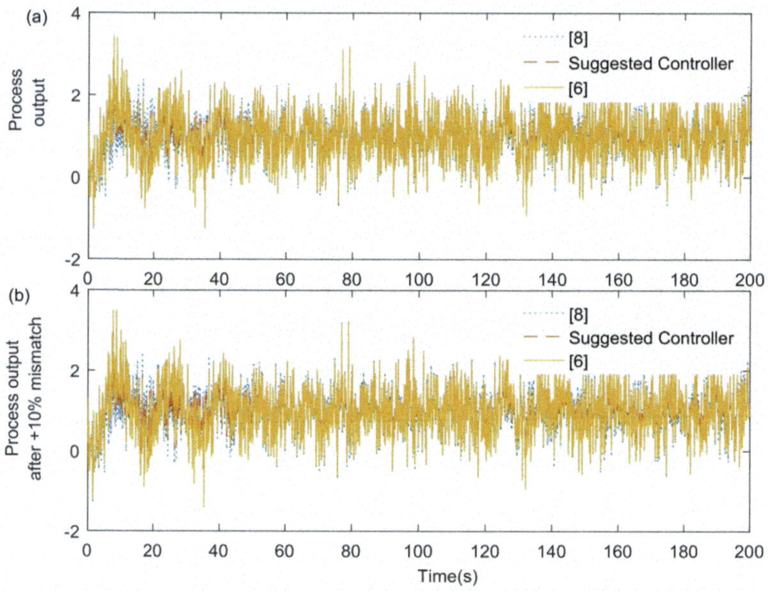

Fig. 2.24 Relative analysis of noise filter for boiler drum level

Remark 2.5 A comparison of the proposed controller with controllers from [6, 8] is conducted using identical parameters. As shown in Fig. 2.24a, b, the existing controllers display problems with overshoot and undershoot. Furthermore, these approaches did not include a noise filter. The loop gain cross-over frequency is $\omega = 0.238$ updated $T_m = 0.019$.

2.3.3.2 Case Study II

The transfer function of the distillation column is given [8].

$$\tilde{P}_1(s) = \frac{(1-0.2s)e^{-0.1s}}{s+1}, \tilde{P}_2(s) = \frac{e^{-0.1s}}{s+1}.$$

After utilization Eqs. (2.24) and (2.25), the noise filter becomes,

$$G_{f_2}(s) = \frac{1}{0.00036s^2 + 0.019s + 1}.$$

Here the loop gain cross-over frequency is $\omega = 0.32$ updated $T_m = 0.023$.

Remarks 2.6 The controller proposed in [8, 9] is linked with the same parameter set. Figure 2.25a, b illustrate that the current controller exhibits overshooting and

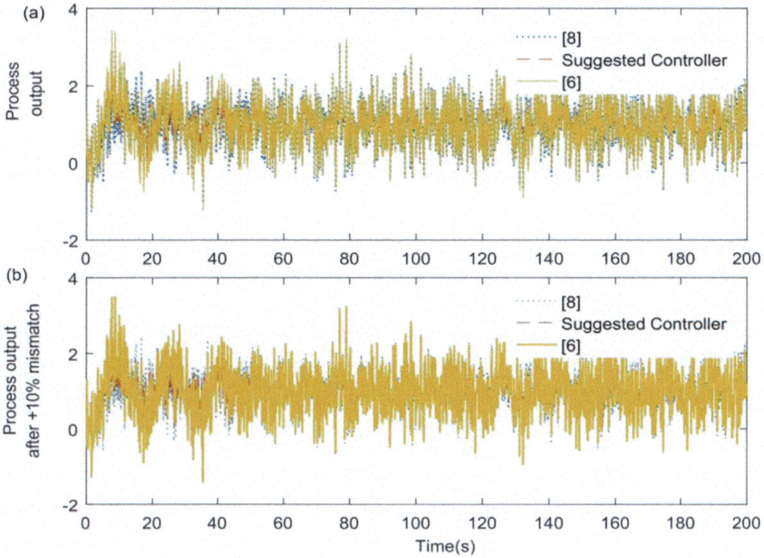

Fig. 2.25 Relative investigation noise filter for the distillation column

undershooting. Their study did not include an extra noise filter. The frequency at which the loop gain crosses over $\omega = 0.32$, updated $T_m = 0.023$.

2.4 Summary

This chapter discusses modifications to the outer loop of the series cascade control scheme by incorporating a compensator to address dead time and inverse response. Since the dead time is relatively small, the supplementary dead-time compensator can be omitted. This topic is detailed in Chap. 3.

2.5 Some Informative Views of Chapter in Post-Mortem Form

Question-1: How by adding compensators, compensate for the inverse response behavior and large dead time of the process?

Answer: *Adding a dead-time compensator and an inverse response compensator to a control system can effectively address the challenges posed by inverse response behavior and large dead times in the process dynamics. Integrating both compensators into the controller allows for a holistic approach to managing process* *dynamics. The inverse response compensator handles the immediate response issues due to inverse behavior, while the dead-time compensator anticipates and mitigates delays in the process. Adding a dead-time compensator and an inverse response compensator to the controller provides a comprehensive solution to manage the challenges associated with inverse response behavior and large dead times in industrial processes. These compensators work together to optimize control performance, stability, and robustness.*

Question-2 Explain the stability analysis of the fractional-order system.

Answer: *Stability analysis of fractional-order systems involves assessing the conditions under which the system remains stable or unstable. Unlike integer-order systems, where stability is typically analyzed using methods like root locus, Nyquist criterion, or Routh-Hurwitz criterion, fractional-order systems require specialized techniques due to their non-integer differentiation and integration orders.*

Riemann surface principle: *The Riemann surface principle is a powerful tool in control theory for analyzing stability and designing controllers for systems with complex and multi-valued dynamics. It extends traditional stability analysis techniques to encompass the nuances of fractional-order systems and other complex transfer functions.*

Question-3 What is the purpose of robustness analysis?

Answer: *Robustness analysis in controller design is critical for ensuring reliable and stable operation of control systems across diverse operating conditions and uncertainties. It involves rigorous assessment of sensitivity, stability margins, and performance metrics to achieve optimal system performance.*

Question-4 Can you explain the importance of noise filters in control systems and how they contribute to system stability and performance?

Answer: ***Impact on System Stability***: *Noise filters are essential components in control systems because they help attenuate unwanted signals (noise) that can interfere with the proper functioning of the system. Noise can originate from various sources such as electromagnetic interference (EMI), power supply fluctuations, or sensor inaccuracies. If not filtered out, this noise can corrupt the control signals and lead to erratic system behavior.*

Performance Enhancement: *The implementation of noise filters in control systems leads to significant performance enhancements by improving signal quality, reducing interference, minimizing errors, enhancing stability, and enabling faster responses. These benefits collectively contribute to the system's ability to achieve precise control and reliable operation in var.*

Question-5 Describe how you would design a filter to handle measurement noise in a control system. What considerations would you take into account, and what types of filters might you use in industrial and technological applications.

Answer:

Understand the Problem

Identify the Nature of Noise: Determine the characteristics of the measurement noise. Is it white noise, colored noise, or some other type? Understanding this will guide your choice of filter.

Understand System Requirements: Clarify the performance requirements of the control system, including how much noise reduction is needed and how it will impact system stability and performance.

Choose the Filter Type

Kalman Filter:

Purpose: To provide an optimal estimate of the state of the system considering both the process model and measurement noise.

Design Considerations: Requires defining the state-space model, process noise covariance, and measurement noise covariance. It dynamically adjusts estimates based on new measurements.

Adaptive Filter:

Purpose: To adaptively adjust filter parameters to handle changing noise characteristics.

Design Considerations: Includes filters like LMS (Least Mean Squares) or RLS (Recursive Least Squares). These filters adjust their coefficients in real time based on incoming data.

Design Considerations

Frequency Response: Ensure that the filter's frequency response is appropriate for the noise and signal frequency ranges. The filter should effectively attenuate noise without distorting the desired signal.

Latency and Computational Complexity: Evaluate the trade-offs between filter complexity and computational efficiency. Ensure that the filter does not introduce unacceptable delays or require excessive computation.

Robustness: Design the filter to be robust to variations in noise characteristics and system dynamics. This might involve considering worst-case scenarios or varying operational conditions.

Stability and Performance: Ensure that the filter design does not adversely affect the overall stability and performance of the control system. Check for potential issues like phase lag or resonance.

Example Application

Example Scenario: Suppose you have a control system with a noisy sensor measuring temperature, where the noise is primarily white. A low-pass filter with a cutoff frequency set based on the noise and signal characteristics might be a good starting point. If the noise characteristics vary over time or are not well understood, a Kalman filter might provide better results by dynamically adjusting to the changing conditions.

Prototype and Test: Implement the chosen filter design and test it in a simulation or real-world scenario to ensure it meets the desired performance criteria.

Iterate: Based on testing results, refine the filter design if necessary to improve performance or address any issues encountered.

In control systems, how do you handle measurement noise, and what strategies or techniques would you use to ensure accurate system performance?

Understanding Measurement Noise

Nature of Noise: Explain that measurement noise can be random variations that distort the true signal being measured. This noise can be white noise (uniform across frequencies) or colored noise (varying with frequency).

Impact on Control Systems: Discuss how measurement noise can affect the accuracy and performance of a control system by causing incorrect feedback, leading to suboptimal control actions and potentially destabilizing the system.

Strategies for Handling Measurement NoiseFiltering:

Low-Pass Filters:

- *Purpose: To attenuate high-frequency noise while allowing the desired signal to pass.*
- *Design Considerations: Choose an appropriate cutoff frequency to balance between noise reduction and signal fidelity. Examples include moving average filters, Butterworth filters, or Chebyshev filters.*

Kalman Filter:

- *Purpose: To estimate the state of a system optimally by combining predictions from a model and actual measurements, considering both process and measurement noise.*

- *Design Considerations: Requires accurate system modeling and knowledge of noise characteristics. The Kalman filter adjusts its estimates based on the predicted state and new measurements.*

Wiener Filter:

- *Purpose: To minimize the mean square error between the desired signal and the filtered output, based on statistical properties of the signal and noise.*
- *Design Considerations: Requires knowledge of the power spectral densities of both the signal and the noise.*

Adaptive Filters:

- *Purpose: To adjust filter parameters in real time to cope with changing noise characteristics.*
- *Design Considerations: Filters like LMS (Least Mean Squares) or RLS (Recursive Least Squares) adapt their coefficients dynamically based on the input signal.*

Sensor Calibration and Improvement:

- *Purpose: To reduce the measurement noise by improving sensor accuracy and precision.*
- *Methods: Regular calibration, using higher-quality sensors, or employing sensor fusion techniques to combine multiple sensor readings for more accurate measurement.*

Signal Averaging:

- *Purpose: To reduce noise by averaging multiple measurements to smooth out random fluctuations.*
- *Implementation: Implement averaging algorithms that combine several measurements to reduce the impact of noise.*

Model-Based Approaches:

- *Purpose: To use a system model to predict and correct for measurement noise.*
- *Techniques: Incorporate system dynamics and noise characteristics into the control algorithm to mitigate the effects of noise.*

Implementation and Testing:

- *Prototype: Develop and test the chosen noise handling strategy in a controlled environment or simulation.*
- *Evaluate Performance: Assess the effectiveness of the noise reduction approach by measuring its impact on system accuracy and stability.*
- *Iterate: Refine the approach based on testing results and practical performance to ensure it meets the required specifications and operational conditions.*

References

1. Stephanopoulos G (1984) Chemical process control. Englewood Cliffs, New Jersey, Prentice Hall
2. Begum KG, Rao AS, Radhakrishnala TK (2017) Enhanced IMC-based PID controller design for non-minimum phase (NMP) integrating processes with time delays. ISA Trans 68:223–234
3. Nadh A, Samuel J, Sharma A (2017) Taylor series approximation of self-interference channel in full-duplex radios. IEEE Trans Wirel Commun 16(7):4304–4316
4. Yao R, Sun K, Qiu F (2019) Vectorized efficient computation of Padé approximation for semi-an simulation of large-scale power systems. IEEE Trans Power Syst 34(5):3957–3959
5. Gupta S, Singh A, Deb D, Ozana S (2021) Kalman filter and variants for estimation in 2 DOF serial flexible link and joint using fractional order PID controller. Appl Sci 11(15):6693
6. Deb D, Patel R, Balas VE (2020) A review of control-oriented bioelectrochemical mathematical models of microbial fuel cells. Processes 8(5):583
7. Kundu S, Singh M, Giri AK (2022) Implementation of variable gain controller based improved phase locked loop approach to enhance power quality in an autonomous microgrid. Int J Numer Model Electron Netw Dev Fields 36(5):e3082
8. Ranganayakulu R, Rao AS, Bhasakar B (2020) Analytical design of fractional IMC filter–PID control strategy for performance enhancement of cascade control systems. Int J Syst Sci 51(10):1699–1713
9. Nagarsheth SH, Sharma SN (2020) Control of non-minimum phase system with dead time: a fractional system viewpoint. Int J Syst Sci 51(11):1905–1928
10. Qibing J, Ling Q, Xuewei W, Fei Q (2009) Based on all pole approximations, a new IMC-PID method for the time-delay system. In: Proceedings of international conference on mechatronics and automation. Changchun, China, pp 268–273
11. Kaya I (2020) Integral-proportional derivative tuning for optimal closed-loop responses to control integrating processes with an inverse response. Trans Inst Meas Control 42(16):3123–3134
12. Raja GL, Ali A (2021) Enhanced tuning of Smith predictor-based series cascade control structure for integrating. ISA Trans 114:191–205
13. Segovia VR, Haggulund T, Astrom KJ (2014) Measurement noise filtering for PID controllers. J Process Control 24(4):299–313
14. Goud EC, Rao AS (2020) Design of noise filter for integrating time delay process. Chem Prod Process Model 15(2):20190056
15. Joshi D, Deb D, Giri AK (2024) MRAS disturbance observer-based sensorless field-oriented backstepping control of BLDC motor drive. In: Electrical engineering. Springer, Berlin. https://doi.org/10.1007/s00202-024-02378
16. Kundu S, Singh M, Giri AK (2022) Implementation of in-loop filter based DSC-FLL control approach for power quality improvement in isolated microgrid. In: Proceedings of 2022 IEEE 31st international symposium on industrial electronics (ISIE). USA. https://doi.org/10.1109/ISIE51582.2022.9831712
17. Giri AK, Arya SR, Maurya R (2020) Hybrid order generalized integrator based control for VSC to improve the PMSG operation in isolated mode. In: The proceedings of 2020 first international conference on power, control and computing technologies (ICPC2T). India. https://doi.org/10.1109/ICPC2T48082.2020.9071498
18. Kundu S, Singh M, Giri AK (2024) Synchronization and control of WECS-SPV-BSS-based distributed generation system using ICCF-PLL control approach. Electr Power Syst Res 226:109919
19. Giri AK, Arya SR, Guerrero JM, Kumar S, Power quality: infrastructures and control. Springer Nature, Singapore. ISBN: 978-981-19-7956-9
20. Giri AK, Arya SR, Vinnikov D, Distributed energy systems design, modeling, and control, vol 1. CRC Press, pp 1–308. ISBN: 9781032134246

21. Arya SR, Patel MM, Alam SJ, Srikakolapu J, Giri AK, Chitti Babu B (2021) Classical control algorithms for permanent magnet synchronous generator driven by diesel engine for power quality. Int J Circuit Theory Appl 49(3):576–601
22. Kundu S, Giri AK, Singh M, Kadiyan S, Adaptive control strategy for isolated renewable energy-based generation system with intermittency. In: Electrical engineering. Springer, Berlin. https://doi.org/10.1007/s00202-024-02454-0
23. Meena DC, Singh M, Giri AK (2021) Leaky-momentum control algorithm for voltage and frequency control of three-phase SEIG feeding isolated load. J Eng Res ICARI Spec 109:120
24. Meena DC, Singh M, Giri AK (2021) A modified NLMS control algorithm for coordinated operation in three-phase wind-energy conversion system. Int J Renew Energy Res (IJRER) 11(4):1621–1629
25. Dalwadi N, Deb D, Ozana S (2023) Performance evaluation of RW-quadrotor and Bi-quadrotor for payload delivery. IET Intel Transp Syst 17(11):2221–2236
26. Dalwadi N, Deb D, Ozana S, Adaptive hybrid control of quadrotor drones. 461:1–174. ISBN: 978-981-19-9743-3

Chapter 3
Advancements in the Cascade Control Scheme

3.1 An Improvement in a Series Cascade Control Scheme

Systems with Non-minimum phase behavior and dubious delay are universal in process industries. The non-minimum phase (NMP) characteristics contemporary a substantial task in designing controllers, as they impose constraints on the performance of the feedback system.

Vanavil et al. [1] presented an innovative IMC–PID controller scheme that unites a lead-lag filter to stabilize integrating unstable systems with dead time. Another IMC approach was suggested for integrating processes, where the integrator is approximated using a lag filter with a prominent time constant [2, 3]. A systematic cascade control system with a fractional filter in the IMC framework is sketched [4].

This chapter introduces a new control configuration using a higher-order fractional IMC filter unified with a higher-order Pade approximation. These higher-order filters can provide greater attenuation and on the other hand, higher-order approximations can yield more exact information about stability margins and robustness of the control system.

3.2 Series Cascade Control Scheme with Inverse Response Compensator

In this suggested method inverse response compensator is combined with the outer loop controller of series cascade control (Fig. 3.1) in a fractional higher-order filter in the IMC paradigm. The details of the inner loop controller's transfer function have been elaborated on earlier.

Inner loop controller: The outline of the inner loop controller is demonstrated [1].

M. Yadav et al., *Control Strategies for Non-minimum Phase Systems*,
Studies in Infrastructure and Control, https://doi.org/10.1007/978-981-96-2351-8_3

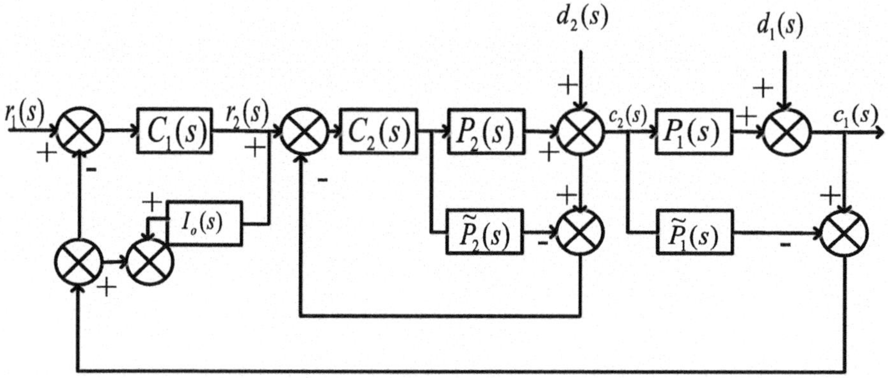

Fig. 3.1 Suggested series cascade control outline

$$C_2(s) = \frac{f(s)}{\tilde{P}_2^-(s)}.$$

Outer loop controller design: Primarily, we consequent the overall process model to sketch the outer loop controller.

The outer loop controller in Fig. 3.2 reflects adjustments made after reconsidering the inverse response compensator. Here we deduce the controller transfer function [5].

$$C''(s) = \frac{C_1(s)}{1 + C_1(s)I_o(s)}. \tag{3.1}$$

The global IMC control structure can be seen in Fig. 3.3 [6].

The classical structure of the feedback controller is elucidated in Fig. 3.3,

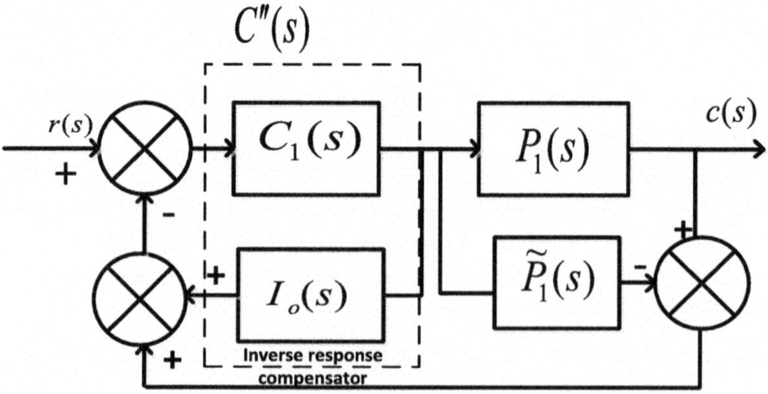

Fig. 3.2 Basic recommended scheme in IMC framework

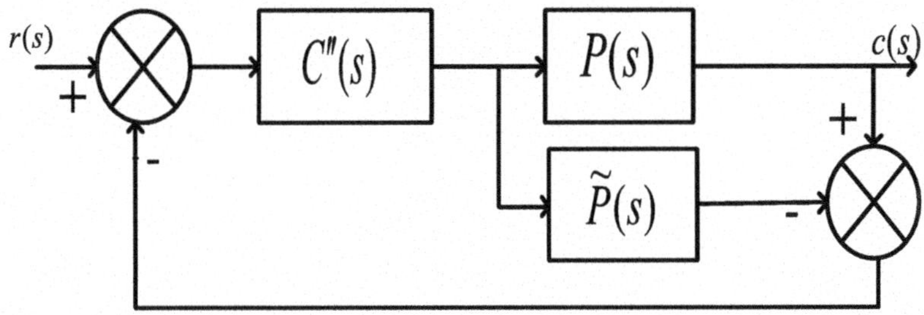

Fig. 3.3 Traditional IMC scheme

$$C(s) = \frac{c(s)}{r(s)} = \frac{C''(s)}{1 - C''(s)\tilde{P}(s)}. \tag{3.2}$$

To implement IMC, the process model must be factorized into two parts, i.e., invertible $\tilde{P}^{+}(s)$ and non-invertible parts $\tilde{P}^{-}(s)$. The IMC controller comprises a filter and the reciprocal of. The term $\tilde{P}^{-}(s)$ infers the minimum phase property of the system. The overall controller transfer function is achieved as,

$$C_{proposed}(s) = \frac{C_1(s)}{1 - C_1(s)\tilde{P}(s) - C_1(s)I_o(s)}. \tag{3.3}$$

Furthermore, $I_o(s) = 2as\tilde{P}^{-}(s)$ shows an inverse response compensator [7]. Subsequently substituting the values in Eq. (3.3) the controller transfer function becomes,

$$C_{propsoed}(s) = \frac{\tilde{P}^{+}(s)}{\tilde{P}(s)[(f_2(s))^{-1} - \tilde{P}^{+}(s) - I_o(s)]}. \tag{3.4}$$

Sketch of controllers for cascade loop: FOPTD system in the inner loop and Integer order plus time delay (IOPDT) system in the outer loop

Inner loop controller outline: Consider a FOPDT model,

$$\tilde{P}_2 = \frac{K_2'}{Ts + 1} e^{-t_{d_2} s},$$

where K_2 are the gain, $e^{-t_{d_2} s}$ the time delay, and the time constant.

From Eq. (2.7), the inner loop controller is achieved as,

$$C_2(s) = \frac{(Ts + 1)}{K_2'(\eta s + 1)^n}. \tag{3.5}$$

Outer loop controller: To design the controller with the IOPDT system,

$$\tilde{P}_1(s) = \frac{K_1'(1-as)}{s}e^{-t_{d_1}s},$$

Equation (2.10) displays the total plant transfer function, which can be articulated as.

$$\tilde{P}(s) = \frac{K'(1-as)}{s(\eta s+1)}e^{-t_d s},\ t_d = t_{d_1} + t_{d_2}. \tag{3.6}$$

After implementing Eq. (3.6), the transfer function of the outer loop controller is stated as,

$$C_{proposed1}(s) = \frac{\frac{t_d^2}{12}s^2 + \frac{t_d}{2}s + 1}{\frac{t_d^2 a\tau s^3}{12} + \phi^3 s^{2\lambda+2} + 3\phi^2 s^{2\lambda} + s^2\left(\frac{at_d^2}{12} - \frac{t_d^2\tau}{12} - \frac{t_d\tau a}{2}\right)}$$

$$+s\left(\frac{t_d\tau}{2} - \frac{t_d^2}{12} - \frac{t_d a}{2} - a\tau\right) + 3\phi s^\lambda + \left(\frac{t_d}{2} - a - \tau\right)\left(\frac{\phi\tau s}{K_1'} + \frac{1}{K_1's} + \frac{(t_d+\phi)}{K_1'}\right). \tag{3.7}$$

Here fractional filter parameters can be tuned as, [6].

$$\lambda = \frac{\pi - P.M.}{\frac{\pi}{2}} - 1,\ \phi = \frac{1}{(\omega)^{\lambda+1}}. \tag{3.8}$$

τ is the extra degree of freedom that can be tuned via the following expressions [7],

$$\tau = \phi\left[1 - \sqrt{(1-\frac{\lambda}{\phi})^3 e^{-\frac{t_d}{\phi}}}\right].$$

Now first-order fractional IMC filters, i.e., $f_2(s) = \frac{1}{\phi s^{\lambda+1}+1}$ and higher-order Pade approximation are considered for time delay.

$$C_{proposed2}(s) = \frac{\frac{t_d^2}{12}s^2 + \frac{t_d}{2}s + 1}{\frac{at_d^2 s^2}{12} - s(\frac{t_d^2}{2} + \frac{\tau^2}{12}) + \phi s^\lambda + (\frac{t_d}{2} - a)}\left(\frac{\phi s}{K_1'} + \frac{1}{K_1'}\right). \tag{3.9}$$

Design of controllers for cascade loop: FOPDT system in the inner loop, FOPDT system in the outer loop

Inner loop controller design: The inner loop controller transfer function is presented in Eq. (3.5).

$$C_s(s) = \frac{(Ts+1)}{K_2'(\eta s+1)^n}.$$

Outer loop controller design: Consider a FOPDT system,

$$\tilde{P}_1(s) = \frac{K_1'(1-as)}{T_1 s+1} e^{-t_{d_1} s}.$$

After incorporating Eq. (2.10), the outer loop process model takes the form,

$$\tilde{P}(s) = \frac{K'(1-as)}{(T_1 s+1)(\eta s+1)} e^{-t_d s}. \tag{3.10}$$

Now resulted in the implementation of the higher-order fractional IMC filter [6]

$$f_2(s) = \frac{\tau s+1}{(\phi s^{\lambda+1}+1)^2}.$$

Here Eq. (3.6) leads to the outer loop controller transfer function. It evolves into,

$$C_{proposed3}(s) = \frac{\left(\frac{t_d^2}{12}s^2 + \frac{t_d}{2}s+1\right)(T_1 s+1)}{\begin{array}{l}\frac{t_d^2 a\tau s^3}{12} + \phi^2 s^{2\lambda+2} + s^2\left(\frac{t_d^2 a}{12} - \frac{t_d^2\tau}{12} - \frac{a\tau t_d}{2}\right) + s\left(\frac{t_d\tau}{2} - a\tau - \frac{t_d a}{2} - \frac{t_d^2}{12}\right) \\ +2\phi s^{\lambda} + (\frac{t_d}{2} - a - \tau)\end{array}}\left(\frac{(\phi+\tau)}{K_1'} + \frac{1}{K_1' s} + \frac{\tau\phi s}{K_1'}\right) \tag{3.11}$$

Equation (3.6) yields the transfer function of the outer loop controller (Table 3.1).

$$C_{proposd4}(s) = \frac{\frac{t_d^2}{12}s^2 + \frac{t_d}{2}s + 1}{\frac{t_d^2 a}{12}s^2 - s(\frac{t_d^2}{12} + \frac{t_d a}{12}) + \phi s^{\lambda} + (\frac{t_d}{2} - a)} \left(\frac{(\phi+T_1)}{K_1'} + \frac{1}{K_1' s} + \frac{T_1\phi s}{K_1'}\right). \tag{3.12}$$

Table 3.1 Filters and approximations for the suggested structure

Adopted 1	$f_1(s) = \frac{1}{\eta s+1}$(inner loop) and $f_2(s) = \frac{\tau s+1}{(\phi s^{\lambda+1}+1)^3}$ (outer loop) + 2nd order Pade approximation
Adopted 2	$f_1(s) = \frac{1}{\eta s+1}$(inner loop) and $f_2(s) = \frac{1}{(\phi s^{\lambda+1}+1)}$ (outer loop) + 2nd order Pade approximation
Adopted 3	$f_1(s) = \frac{1}{\eta s+1}$(inner loop) and $f_2(s) = \frac{\tau s+1}{(\phi s^{\lambda+1}+1)^2}$(outer loop) + 2nd order Pade approximation
Adopted 4	$f_1(s) = \frac{1}{\eta s+1}$ inner loop) and $f_2(s) = \frac{1}{(\phi s^{\lambda+1}+1)}$ (outer loop) + 2nd order Pade approximation

3.3 Case Studies

The effectiveness of suggested scheme is tested via two benchmark problems.

3.3.1 Case Study I

Singhal et al. [6] offered the transfer function of the boiler drum level system to be as follows:

$$\tilde{P}_1(s) = \frac{(1 - 0.418s)e^{-0.07s}}{s}, \tilde{P}_2(s) = \frac{0.547e^{-0.03s}}{1.06s + 1}.$$

Inner loop controller design: In configuring the inner loop controller using Eq. (3.5),

$$C_2(s) = \frac{(1 + 1.06s)e^{-0.1s}}{0.547(0.3s + 1)}.$$

Outer loop controller design: After pertaining Eq. (3.6), the transfer function of the plant can be reformulated as,

$$\tilde{P}(s) = \frac{(1 - 0.418s)e^{-0.1s}}{s(0.3s + 1)}. \tag{3.13}$$

After utilization Eq. (3.7),

$$C_{proposed1}(s) = \frac{0.008s^2 + 0.05s + 1}{s^{2.90} + 0.0369s^2 - 0.267s + 3s^{0.90} + 3s^{0.45} - 0.568}\left(0.38 + \frac{1}{s} + 0.0336s\right). \tag{3.14}$$

Where the term $\frac{0.008s^2+0.05s+1}{s^{2.90}+0.0369s^2-0.267s+3s^{0.90}+3s^{0.45}-0.568}$ denotes the fractional IMC filter and the term $\left(0.38 + \frac{1}{s} + 0.0336s\right)$ deduces the PID controller. Here Eq. (3.9) the recommended outer loop controller settings become,

$$C_{proposed2}(s) = \frac{0.008s^2 + 0.05s + 1}{0.00068s^2 - 0.021s + s^{0.45} - 0.368}\left(0.38 + \frac{1}{s}\right). \tag{3.15}$$

In Fig. 3.4a, b, it is evident that the existing controller exhibits overshoot and undershoot. The proposed technique shows minimal peak overshoot following disturbances and achieves a moderately faster settling time.

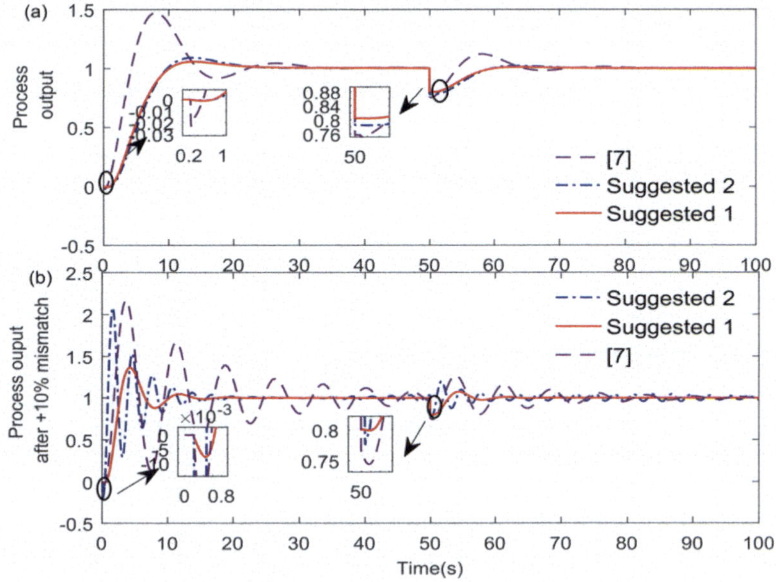

Fig. 3.4 Comparative analysis for boiler drum level system

Refer to Table 3.2 for the Integrated Absolute Error (IAE) and Integrated Squared Error (ISE) values.

Furthermore, the evaluation of stability for the suggested outline is conducted using the Riemann surface, and detailed stability analysis is available in Table 3.3 (Tables 3.4 and 3.5).

The illustration in Fig. 3.5 indicates that all roots extend beyond the Riemann sheet, affirming the stability of the system. The universal sensitivity function is stated as [7],

Table 3.2 Transient and performance indices of boiler drum level system

	Nominal case					+10% mismatch				
Method	Peak O.S. ratio, %	Peak time, (s)	Rise time, (s)	IAE	ISE	Peak O.S. ratio, %	Peak time (s)	Rise time (s)	IAE	ISE
Nagarsheth and Sharma [7]	1.47	8.29	4.26	1.3	1.7	2.16	3.69	1.87	3.4	11.72
Suggested controller 1	1.061	13.37	9.87	0.16	0.027	1.364	4.17	2.82	0.76	0.58
Suggested controller 2	1.094	13.66	9.73	0.651	0.42	2.08	1.67	1.07	2.5	6.25

Table 3.3 Universal expression for the fractional characteristic polynomial

Proposed 1	$f(s)=1+\frac{t_d a\tau s^4}{12}+\phi^3 s^{3\lambda+3}+s^3\left(\frac{t_d^2 a}{12}-\frac{t_d^2\tau}{12}-\frac{\gamma b\lambda}{2}\right)+3\phi^2 s^{2\lambda+2}$ $+s^2\left(\frac{t_d\tau}{2}-\frac{t_d^2}{12}-\frac{t_d a}{2}-a\tau\right)+3\phi s^{\lambda+1}+s\left(\frac{t_d}{2}-a-\tau\right).$
Proposed 2	$f(s)=1+\frac{t_d^2 a s^3}{12}-s^2\left(\frac{t_d^2}{12}+\frac{t_d a}{2}\right)+\phi s^{\lambda+1}+s\left(\frac{t_d}{2}-a\right).$

Table 3.4 Fractional quasi-characteristics polynomial

Proposed 1	$f(s)=1+0.00068s^3-0.021s^2+s^{1.45}-0.368s.$
Proposed 2	$f(s)=1+0.00068s^3-0.021s^2+s^{1.45}+0.367s$

Table 3.5 Mapping of fractional characteristics polynomial

Proposed 1	$f(\alpha^{1/\beta})=1+0.00048\alpha^{4000}+0.0369\alpha^{3000}+\alpha^{2900}-0.267\alpha^{2000}$ $+3\alpha^{1900}+3\alpha^{1450}-0.568\alpha^{1000}.$
Proposed 2	$f(\alpha^{1/\beta})=1+0.00068\alpha^{3000}-0.021\alpha^{2000}+\alpha^{1450}-0.368\alpha^{1000}.$

$$S(s)=\frac{1}{1+\tilde{P}(s)C_{proposed}(s)}. \tag{3.16}$$

The robustness analysis incorporates Eqs. (3.14), (3.15), and (3.16). In Fig. 3.6, you can see the graphical representation of the robustness analysis conducted for the offered controller.

Remarks 3.1 Within this research, we construct a higher-order fractional IMC filter and an inverse response compensator using a series cascade control framework. This method aims to diminish the necessity for a setpoint filter and dead time compensator, and we associate it with [7]. Further same work can be compared with the sliding model and LQR control [8, 9].

3.3.2 *Case Study II*

The distillation column's transfer function is formulated as [7],

$$\tilde{P}_1(s)=\frac{(1-0.2s)e^{-0.1s}}{s+1},\ \tilde{P}_2(s)=\frac{e^{-0.1s}}{s+1}.$$

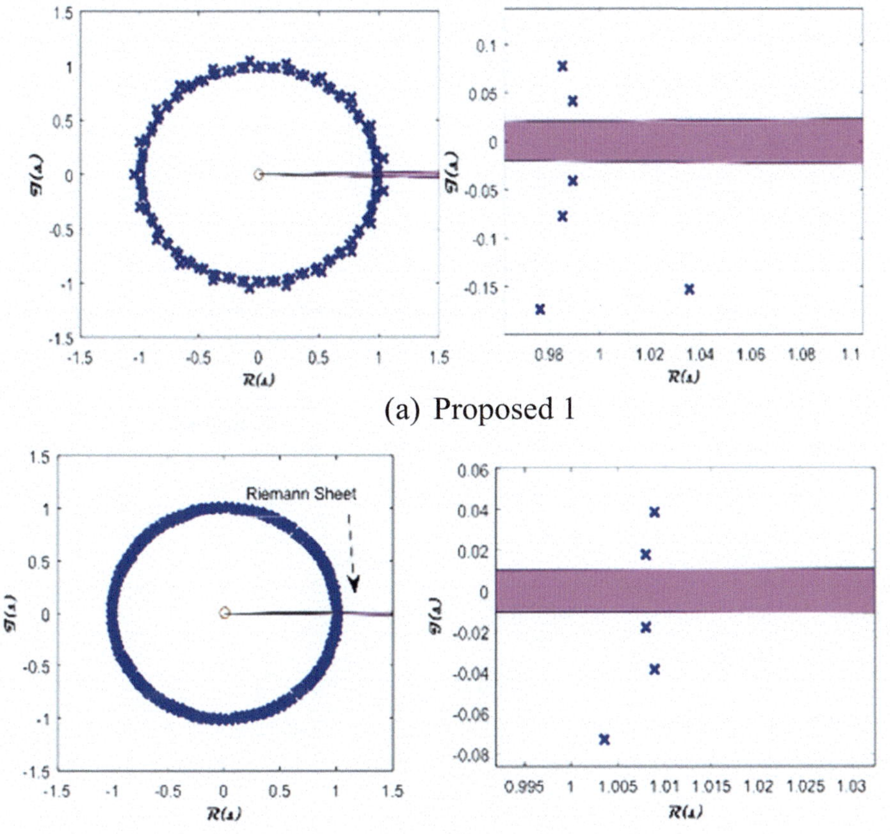

(a) Proposed 1

(b) Proposed 2

Fig. 3.5 Stability analysis for boiler drum level system

Inner loop controller design: After utilizing Eq. (3.5),

$$C_2(s) = \frac{s+1}{(0.35s+1)}.$$

Outer loop controller design: The plant transfer function becomes (Eq. 3.10)

$$\tilde{P}(s) = \frac{(1-0.2s)e^{-0.2s}}{(s+1)(0.35s+1)}. \tag{3.17}$$

Likewise, from Eq. (3.11), the recommended controller settings for the outer loop are achieved as,

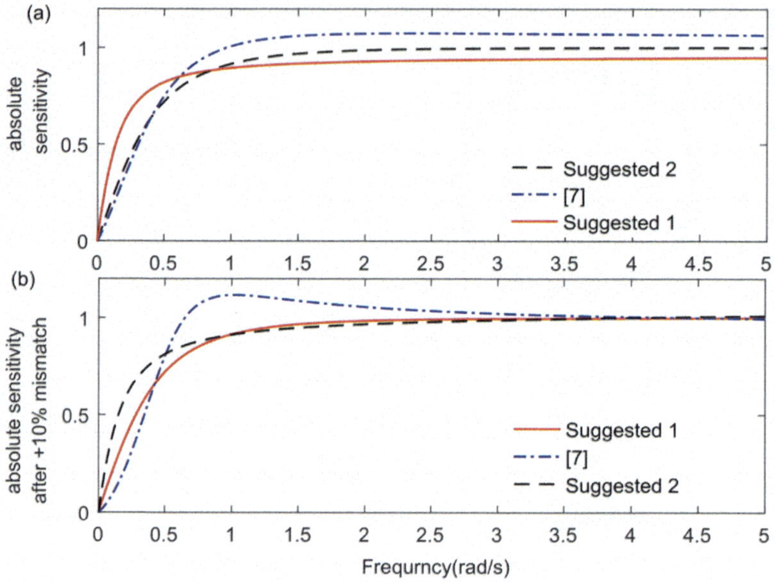

Fig. 3.6 Robustness analysis for boiler drum level system

$$C_{proposed3}(s) = \frac{(0.003s^2 + 0.1s + 1)(s + 1)}{0.000093s^3 + s^{2.56} + 0.00386s^2 - 0.054s + 2s^{0.28} - 0.24} \left(1.14 + 0.14s + \frac{1}{s}\right), \tag{3.18}$$

The term $\frac{(0.003s^2+0.1s+1)(s+1)}{0.000093s^3 + s^{2.56} + 0.00386s^2 - 0.054s + 2s^{0.28} - 0.24}$ indicates fractional IMC filter, while the term $\left(1.14 + 0.14s + \frac{1}{s}\right)$ demonstrates PID controller.

From Eq. (3.12)

$$C_{proposed4}(s) = \frac{0.003s^2 + 0.1s + 1}{0.0006s^2 - 0.023s + s^{0.28} - 0.1} \left(1.35 + \frac{1}{s} + 0.35s\right). \tag{3.19}$$

where the term $\frac{0.003s^2+0.1s+1}{0.0006s^2-0.023s+s^{0.28}-0.1}$ specifies a lower-order fractional filter,

While the term validates PID controller. Here $\eta = 0.35$, $a = 0.1$, $\phi = 1$, and $\lambda = 0.28$ with $PM = 65$.

Figure 3.7 offers a relative assessment of the closed-loop performance achieved using the proposed approach. Figure 3.7a, b illustrate substantial improvements in overshoot and undershoot characteristics under nominal and +10% mismatch conditions.

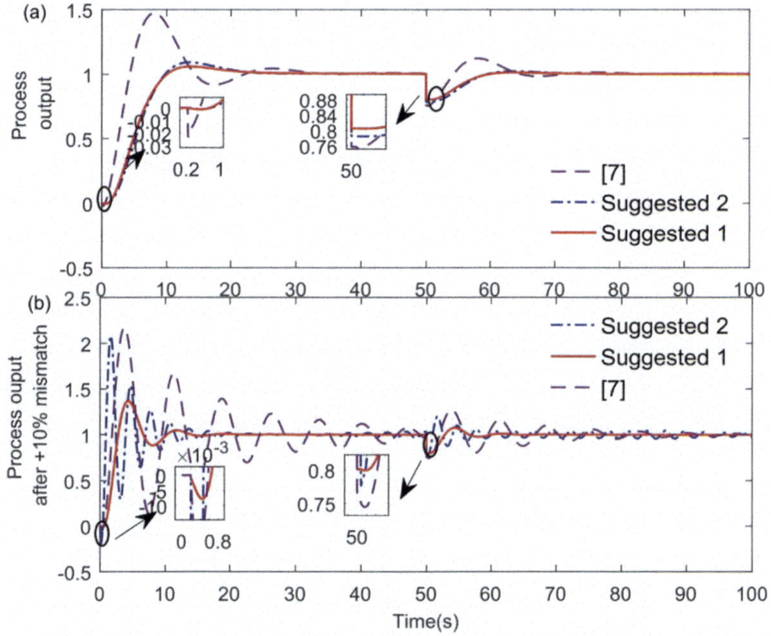

Fig. 3.7 Relative analysis for the distillation column

In Table 3.6, you can find the transient response and performance indices. Furthermore, Table 3.7 outlines the general expression for stability analysis.

The detailed setup of the fractional characteristic polynomial, which is linked to the closed-loop response of the distillation column, is presented in Table 3.8 (Table 3.9).

In both cases presented in Table 3.10, all roots are situated outside the Riemann sheet, affirming the stability of the system.

Table 3.6 Transient and performance indices for the distillation column

Method	Nominal case					+10% mismatch				
	Peak O.S. ratio, %	Peak time (s)	Rise time (s)	IAE	ISE	Peak O.S. ratio, %	Peak time (s)	Rise time (s)	IAE	ISE
[7]	1.5	5.1	2.7	1.8	3.3	1.74	4.6	2.4	1.5	2.3
Suggested controller 1	1.40	9.9	6.18	0.43	0.18	1.46	9.1	5.3	0.34	0.11
Suggested controller 2	1.41	5.8	3.7	1.1	1.3	1.54	6.5	3.7	1.03	1.064

Table 3.7 Generalized expression for stability analysis of the FOPDT system

Proposed 3	$f(s) = 1 + \frac{t_d^2 a\tau s^4}{12} + \phi^2 s^{2\lambda+3} + s^3\left(\frac{t_d^2 a}{12} - \frac{t_d \tau}{12} - \frac{\tau a\lambda}{2}\right) + s^2\left(\frac{t_d}{2} - \frac{t_d^2}{12} - \frac{t_d a}{2} - a\tau\right) + 2\phi s^{\lambda+1} + s\left(\frac{t_d}{2} - a - \tau\right).$
Proposed 4	$f(s) = 1 + \frac{t_d^2 a s^3}{12} + s^2\left(\frac{t_d^2}{12} + \frac{t_d a}{2}\right) + \phi s^{\lambda+1} + s\left(\frac{t_d}{2} - a\right).$

Table 3.8 Fractional characteristic polynomial for the distillation column

Proposed 3	$f(s) = 1 + 0.00048s^4 + 0.0369s^3 + s^{2.90} - 0.267s^2 + 3s^{1.90} + 3s^{1.45} - 0.568s.$
Proposed 4	$f(s) = 1 + 0.00068s^3 - 0.021s^2 + s^{1.45} - 0.368s.$

Table 3.9 Mapping of fractional order characteristics polynomial

Proposed 3	$f(\alpha^{1/\beta}) = 1 + 0.00048\alpha^{4000} + 0.0369\alpha^{3000} + \alpha^{2900} - 0.267\alpha^{2000} + 3\alpha^{1900} + 3\alpha^{1450} - 0.568\alpha^{1000}.$
Proposed 4	$f(\alpha^{1/\beta}) = 1 + 0.00068\alpha^{3000} - 0.021\alpha^{2000} + \alpha^{1450} - 0.368\alpha^{1000}.$

Table 3.10 Stability analysis for the distillation column

Proposed 3	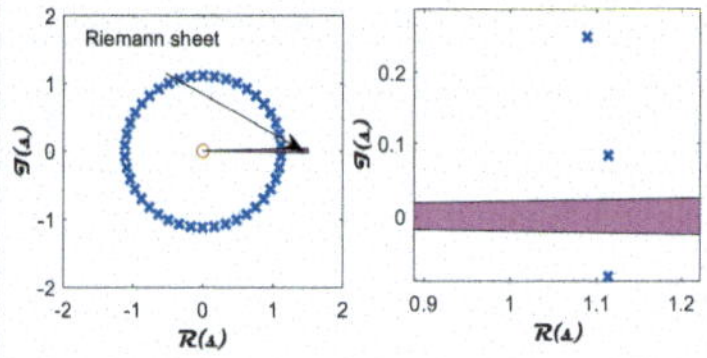
Proposed 4	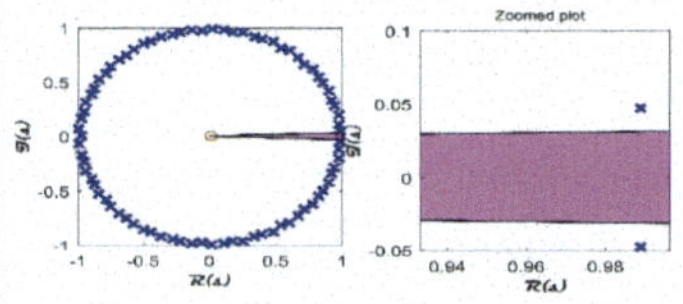

Moreover, to assess the robustness of the proposed controller, the plant and controller transfer function values are specified in Eqs. (3.16), (3.18), and (3.19). Figure 3.8 graphically shows that the suggested controllers exhibit the smallest peak of maximum absolute sensitivity.

Remarks 3.2 The offered controllers are associated using a similar set of parameters as [7]. It has been witnessed that the offered controllers reveal improved performance likened to the existing controller.

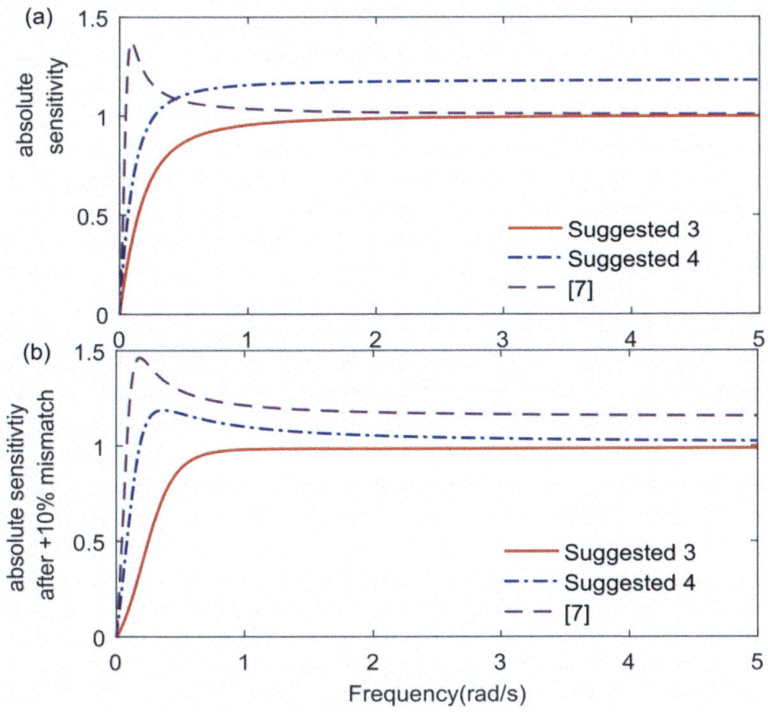

Fig. 3.8 Robustness analysis for the distillation column

3.3.3 Case Study III

The model for a higher-order integrating process in terms of its transfer function is specified as [8].

$$P(s) = \frac{0.5(1 - 0.5s)e^{-0.7s}}{s(0.5s + 1)(0.4s + 1)(0.1s + 1)}.$$

The model for this process is given by [9].

$$P_1(s) = \frac{(1 - 0.4699s)}{s}e^{-0.50s}, P_2(s) = \frac{0.5183}{1.1609s + 1}e^{-0.20s}.$$

Utilizing Eq. (3.5) for the inner loop controller settings results is given as,

$$C_2(s) = \frac{1.1609s + 1}{0.5183(0.95s + 1)}.$$

Equation (3.6) enables the reformation of the overall plant transfer function into:

$$\tilde{P}(s) = \frac{(1 - 0.4699s)e^{-0.7s}}{s(0.95s + 1)}. \tag{3.20}$$

To determine the optimal settings for the outer loop controller in the fractional IMC framework, refer to Eq. (3.7).

$$C_{proposed1}(s) = \frac{0.040s^2 + .35s + 1}{\begin{array}{l}0.017s^3 + 0.168s^2 + 0.6s^{1.65} + 2.16s^{1.30}\\ +0.13s + 2.5s^{0.65} - 1.\end{array}}\left(0.874 + \frac{1}{s} + 1.82s\right), \tag{3.21}$$

where the term $\frac{0.040s^2+0.035s+1}{\begin{array}{l}0.017s^3 + 0.168s^2 + 0.6s^{1.65} + 2.16s^{1.30}\\ +0.13s + 2.5s^{0.65} - 1.\end{array}}$ represents the fractional IMC filter and the term $\left(0.874 + \frac{1}{s} + 1.82s\right)$ interprets the PID controller. From Eq. (3.9),

$$C_{proposed2}(s) = \frac{0.040s^2 + 0.035s + 1}{0.019s^2 - 0.21s + 0.6s^{0.65} - 0.11}\left(0.95 + \frac{1}{s}\right), \tag{3.22}$$

The term signifies the fractional IMC filter and $\left(0.95 + \frac{1}{s}\right)$ interprets the PI controller Here $\phi = 0.92$ $\tau = 0.6$ and $\lambda = 0.65$ with $PM = 39$. The comparative study of the closed-loop response derived from the proposed technique is illustrated in Fig. 3.9. Additionally, Table 3.11 provides the IAE and ISE values.

Stability investigations for the recommended scheme are similarly carried out using the Riemann surface approach. For detailed analysis, please refer to Case Study I, and Table 3.12 presents insights into the stability analysis for a higher-order integrating process. [7].

Moreover, for the robustness analysis, substitute the value of $\tilde{P}(s)$ and $C_{proposed1}(s)$, $C_{proposed2}(s)$ from Eqs. (3.20), (3.21) and (3.22) into (3.16).

The robustness analysis for the proposed controller is illustrated in Fig. 3.10. This analysis shows that the proposed controllers experience a reduced increase in maximum sensitivity in both the nominal case and when faced with a 10% model mismatch.

Remarks 3.3 The efficacy of the suggested controller was assessed by comparing it with the system outlined in [8]. The comparison shows that the prior method employed a single-loop feedback control without an inverse response compensator, while the proposed controllers offer enhanced performance.

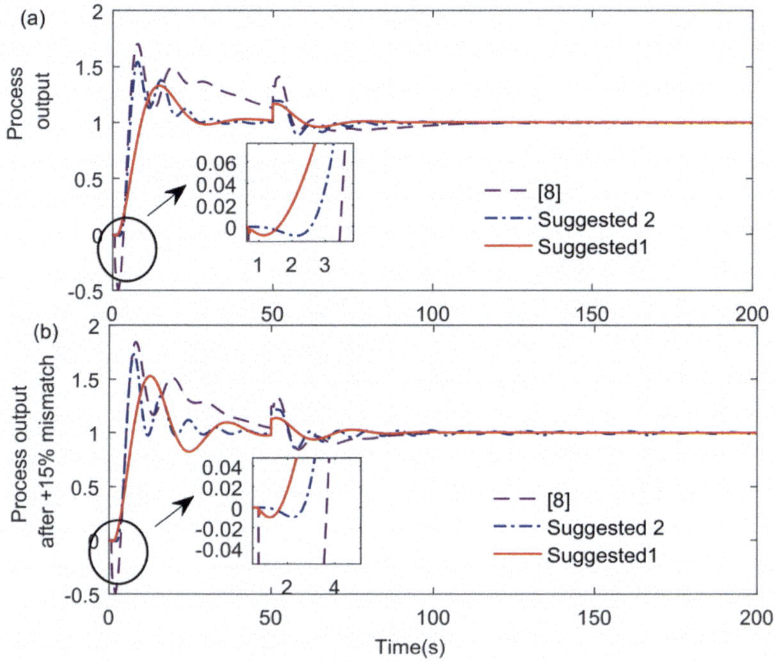

Fig. 3.9 Relative investigation for higher-order integrating process

Table 3.11 Transient and performance indices for higher-order integrating process

	Nominal case					+10% mismatch				
Method	Peak O.S.%	Peak time, s	Rise times	IAE	ISE	Peak O.S. %	Peak time, s	Rise time, s	IAE	ISE
[8]	1.7	7.9	5.2	8.7	76	1.84	8.15	5.4	6.6	44.5
Suggested 1	1.33	14.6	8.5	0.01	0.02	1.53	12.56	7.88	0.16	0.02
Suggested 2	1.54	7.7	5.6	0.79	0.62	1.72	7.9	5.5	0.53	0.28

3.3.4 Case Study IV

We will perform an in-depth analysis of the behavior of a third-order process with both inverse response dynamics and a time delay [6].

$$\tilde{P}(s) = \frac{(1 - 2s)e^{-0.5s}}{(2s + 1)(s + 1)(0.5s + 1)}.$$

Table 3.12 Stability investigation of higher-order integrating process

Proposed 1	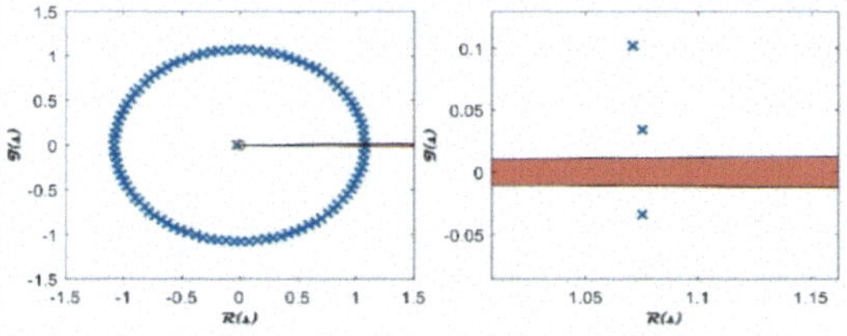
Proposed 2	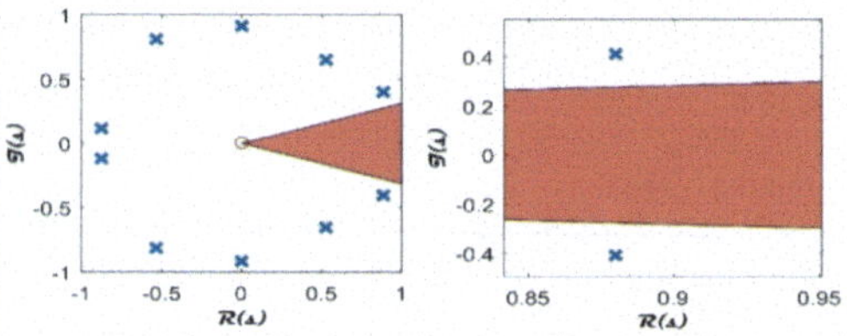

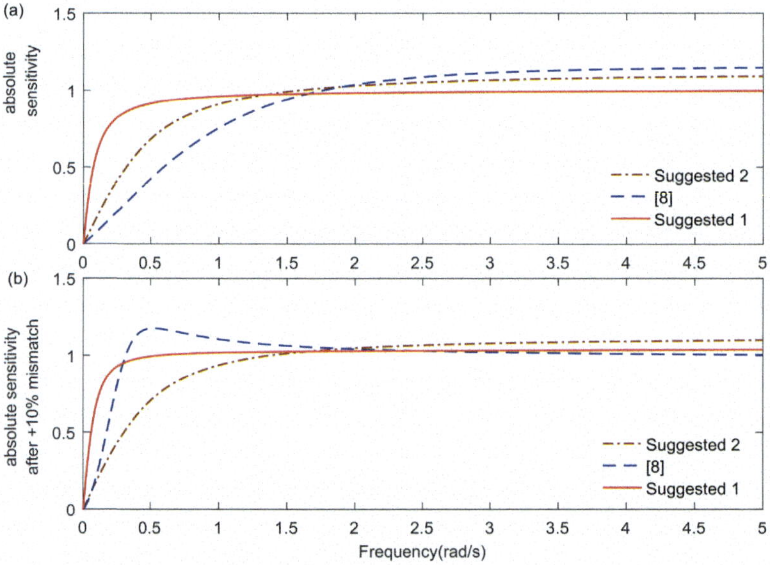

Fig. 3.10 Robustness analysis for higher-order integrating process

Since precise data on the actual process was not available, the model for the controller design was based on the process reaction curve.

$$P_1(s) = \frac{(1 - 1.62s)}{2.71s + 1}e^{-0.4s}, P_2(s) = \frac{0.94}{1.50s + 1}e^{-0.2s}.$$

To establish the inner loop controller settings with Eq. (3.5),

$$C_2(s) = \frac{1.50s + 1}{0.94(0.6s + 1)}$$

Upon using Eq. (3.10), the overall plant transfer function can be redefined as

$$\tilde{P}(s) = \frac{(1 - 1.62s)e^{-0.6s}}{(0.6s + 1)(2.71s + 1)}. \tag{3.23}$$

For the recommended settings of the outer loop controller in the fractional IMC framework, please refer to Eq. (3.11).

$$C_{proposed3}(s) = \frac{(0.03s^2 + .03s + 1)(2.71s + 1)}{0.0012s^3 + 0.5s^{2.84} + 0.7216s^2 + 0.385s + s^{0.42} - 1.01.} \left(1.19 + \frac{1}{s} + 0.35s\right), \tag{3.24}$$

where the term $\frac{(0.030s^2+.03s+1)(2.71s+1)}{0.0012s^3+0.5s^{2.84}+0.7216s^2+0.385s+s^{0.42}-1.01.}$ represents the fractional comparison of the closed-loop response, based on the proposed technique, is presented.

IMC filter and the term $\left(1.19 + \frac{1}{s} + 0.35s\right)$ interprets the PID controller.

$$C_{proposed4}(s) = \frac{0.030s^2 + .03s + 1}{0.048s^2 - 0.11s + 0.5s^{0.42} - 0.32} \left(3.21 + \frac{1}{s} + 0.32\right), \tag{3.25}$$

The term represents the fractional IMC filter and $\left(3.21 + \frac{1}{s} + 0.32\right)$ interprets the PI Controller. Here $\eta = 0.5$ $\phi = 0.59$ $\tau = 0.5$ and $\lambda = 0.42$ with $PM = 52$. In Fig. 3.11.

The transient response and IAE values shown in Table 3.13 confirm the effectiveness of the proposed controller. Even after accounting for a 10% mismatch in process parameters, the controller delivered impressive performance.

The relative positioning of the roots of the natural degree quasi-characteristic polynomial with respect to the Riemann sheet provides insight into system stability. Roots located within the Riemann sheet suggest instability, while roots outside indicate stability, as detailed in Table 3.14.

To examine the robustness of proposed controllers, substitute the value of $\tilde{P}(s)$ and $C_{proposed3}(s)$, $C_{proposed4}(s)$ from Eqs. (3.23), (3.24), and (3.25) into (3.16). Figure 3.12 depicts that the proposed controllers achieve a lower peak in terms of maximum absolute sensitivity.

Remarks 3.4 In contrast to the single-loop feedback method used in [10], this study applies a higher-order fractional IMC filter along with an inverse response compensator and series cascade control to manage setpoint filtering and counteract dead time.

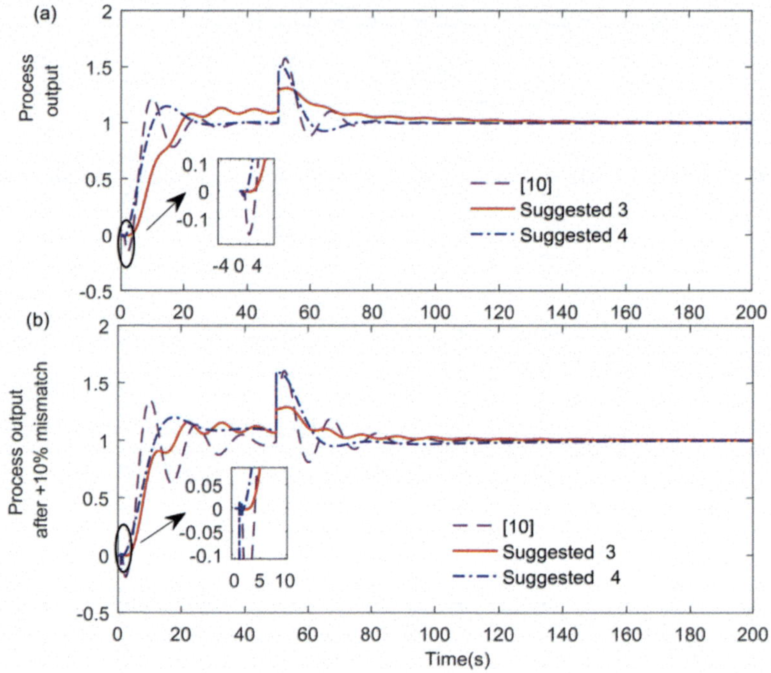

Fig. 3.11 Comparative analysis for third-order process

Table 3.14 Stability analysis of the third-order process

Proposed 3	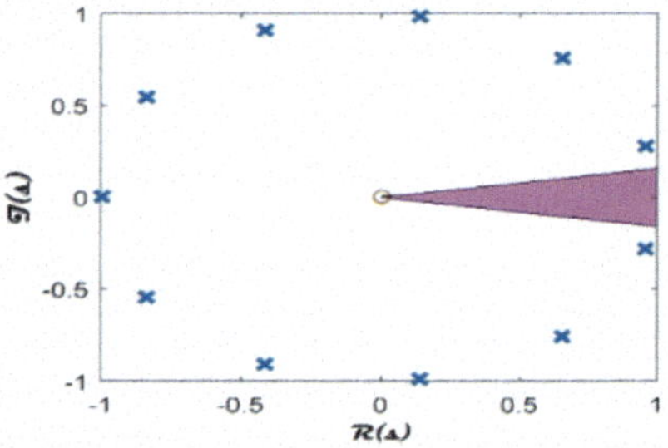
Proposed 4	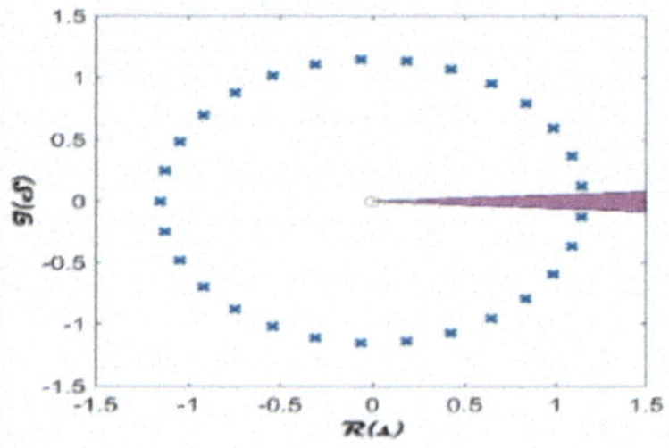

Table 3.13 Transient and performance indices for the third-order process

Method	Nominal case					+10% mismatch				
	Peak O.S %	Peak time, s	Rise time, s	IAE	ISE	Peak O.S. %	Peak time, s	Rise time, s	IAE	ISE
[10]	1.22	9.8	7.5	3.1	9.6	1.35	10.0	7.5	3.3	11.2
Suggested 1	1.13	31.3	19.4	0.16	0.02	1.16	21.8	18.41	0.20	0.04
Suggested 2	1.14	14.2	9.8	2.1	4.7	1.20	17.4	11.75	3.0	9.0

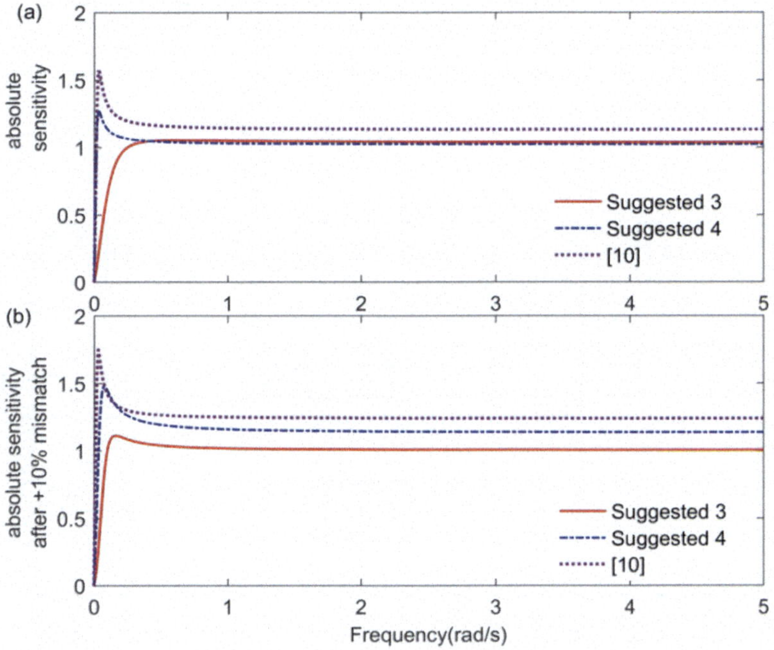

Fig. 3.12 Robustness analysis for third-order process

3.4 Control Structure with Derivative Filter

The outer loop controller design in the IMC framework takes into account the impact of the inverse response compensator as well as the fractional filter.

The enhanced performance and preserved robustness of this combination are demonstrated in Fig. 3.13. Figure 3.14 provides a depiction of the simplified structure of the proposed controller in the IMC paradigm.

The transfer function for the proposed structure can be expressed as follows:

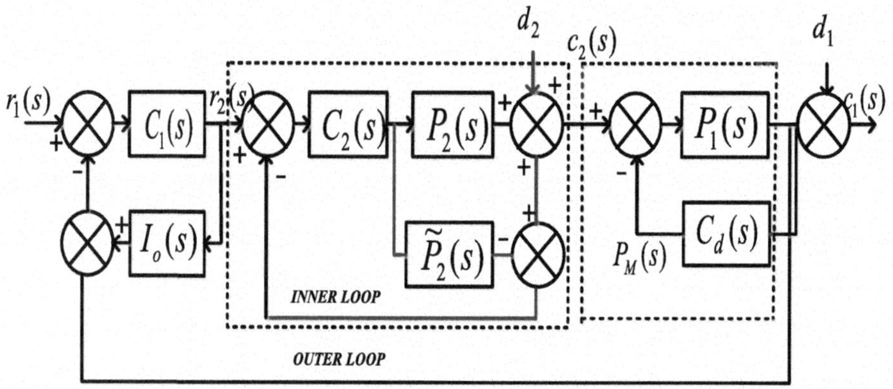

Fig. 3.13 Proposed control structure

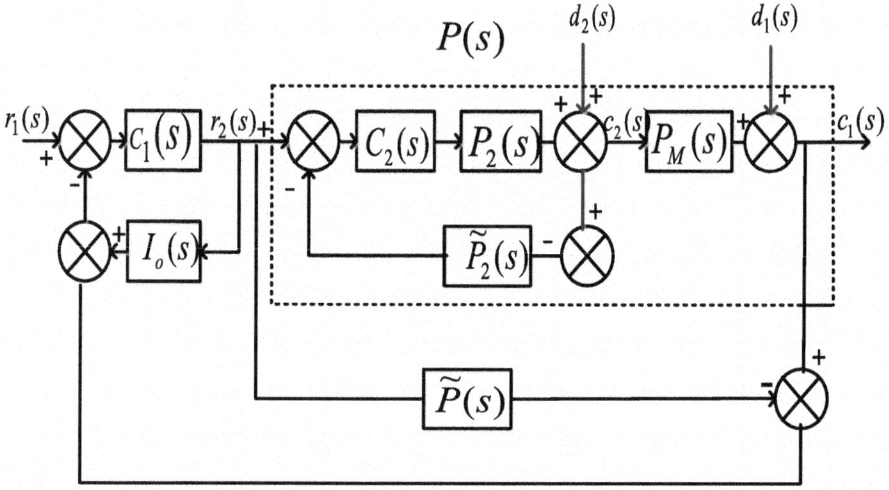

Fig. 3.14 Proposed control strategy in a simplified way

$$
\begin{aligned}
c_1(s) = \tilde{P}_2 s)C_2 s)P_M(s)\{C_1(s)I_o(s)\}r_1(s) + (1 - \tilde{P}_2 s)C_2(s))\\
(1 - P_M(s)C_2(s)\tilde{P}_2(s)\{C_1(s)I_o(s)d_2 +\\
((1 - \tilde{P}_2(s)C_2(s)P_M(s)C_1(s)I_o(s))d_1.
\end{aligned}
$$

3.5 Controller Design

Consider two integrating plants with non-minimum phase zeros [7]. Consequently, the feedback derivative controller (C_d) is defined as

$$C_d(s) = \frac{T_d(s)}{1 + \psi T_d s}$$

ψ is the derivative filter constant, which can vary from 0.01 to 0.1. Now we are using case 1 for controller design.

Sketch of controllers for cascade loop: First order system in the inner loop and Integer order plus time delay (IOPDT) system in the outer loop

Inner loop controller design: The traditional IMC controller designs the inner loop controller [7].

$$\tilde{P}_2(s) = \frac{K_2'}{Ts + 1}$$

In the IMC paradigm, the process model is segregated into two components, i.e., minimum phase system. $\tilde{P}_2^-(s) = \frac{K_2'}{Ts+1}$ and non-minimum phase system $\tilde{P}_2^+(s) = 1$. The generalized IMC controller $C_2(s) = \frac{f(s)}{\tilde{P}_2^-(s)}$ has been used. As a result, the inner loop controller can be given as

$$C_2(s) = \frac{(Ts + 1)}{K_2'(1 + \eta s)}. \tag{3.26}$$

Here T and K_2' are the time constant and proportional gain.

Outer loop controller design: From Fig. 3.13, the feedback derivative controller is embedded with the outer loop process model. The generalized expression $P_M(s)$ can be written as [6].

$$P_M(s) = \frac{\tilde{P}(s)}{1 + \tilde{P}(s)T_d(s)} \tag{3.27}$$

Now consider the outer loop process model

$$\tilde{P}_1(s) = \frac{K_1'(1 - as)}{s}$$

After utilization, Eq. (3.27) $P_M(s)$ becomes,

$$P_M(s) = \frac{K_1'(1 - as)}{s\left(1 + K_1'T_d - aK_1'T_d s\right)} \tag{3.28}$$

To guarantee the stability of the closed-loop system, all roots of the characteristic polynomial must lie in the left half of the root locus plane. This implies that the choice of T_d should be allowed the system roots on the left-hand side. As a result, the characteristic polynomial can be written as,

$$s[(1 + k_1 T_d) - (bk_1 T_d s)] = 0 \tag{3.29}$$

According to R-H criteria $-K_1' a T_d > 0\&\ (1 + K_1' T_d) > 0$. After combining both conditions, the range for selection T_d becomes,

$$\frac{-1}{K_1'} < T_d < 0 \tag{3.30}$$

From Fig. (3.14), the generalized overall process transfer function becomes,

$$\tilde{P}(s) = M(s).P_M(s) \tag{3.31}$$

The overall plant transfer function for this case can be determined from Eq. (3.31) and is represented as:

$$\tilde{P}(s) = \frac{K_1'(1 - as)}{(\eta s + 1)(s + K_1' T_d s - a K_1' T_d s^2)} \tag{3.32}$$

Now simplifying Eq. (3.32) then it becomes,

$$P(s) = \frac{K_1'(1 - as)}{[-aK_1' T_d \eta s^3 + s^2(\eta + K_1' \eta T_d - a K_1' T_d) + s(1 + K_1' T_d)]} \tag{3.33}$$

For simplicity in controller design, the higher-order transfer function can be converted into first order via Ident toolbox in MATLAB. It can be transcribed as,

$$\tilde{P}(s) = \frac{K_1'(1 - as)}{[bs + d]} \tag{3.34}$$

In Eq. (3.34), non-minimum phase zero is present, which limits limiting gain margin and shows inverse response behavior. As a result, an inverse response compensator $I_o(s)$ is embedded with the fractional-filter-based IMC controller $C_1(s)$. Figure 3.14 shows the fractional-filter-based IMC controller after the accountability of the inverse response compensator [7].

From Fig. 3.15

$$C(s) = \frac{C_1(s)}{1 + C_1(s) I_o(s)}. \tag{3.35}$$

The controller transfer function $C(s)$ can be obtained from the IMC approach [7]

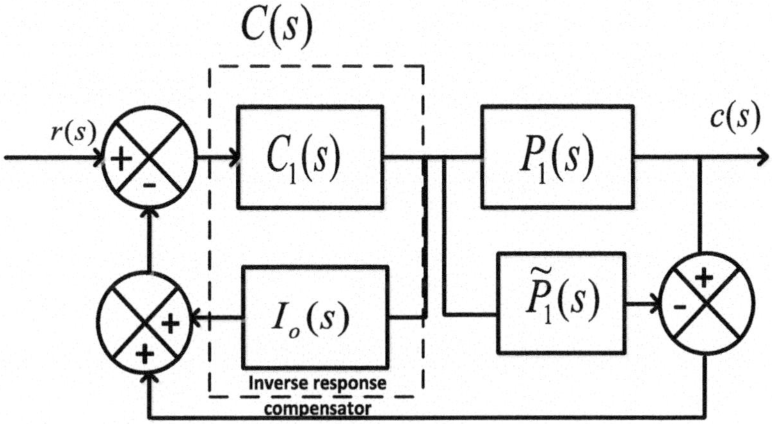

Fig. 3.15 Fractional-order IMC controller with IRC

$$\frac{c(s)}{r(s)} = \frac{\hat{C}(s)}{1 - \hat{C}(s)\tilde{P}(s)} \tag{3.36}$$

Here IMC controller can be written as $\hat{C}(s) = \frac{f(s)}{\tilde{P}^{-}(s)}$. Subsequently, utilization of Eqs. (3.35) and (3.36) with the notion of an equivalent system, the controller transfer function becomes

$$C_1(s) = \frac{\hat{C}(s)}{1 - \hat{C}(s)\tilde{P}(s) - \hat{C}(s)I_o(s)}. \tag{3.37}$$

After utilization, simply Eq. (3.37), the controller transfer function can be reformed as

$$C_1(s) = \frac{\tilde{P}^{+}(s)}{\tilde{P}(s)\big((1+\phi s)^{r} - \tilde{P}^{+}(s)\big) - I_o(s)}. \tag{3.38}$$

The controller transfer function is now designed via Eq. (3.38) after the accountability of the fractional filter and inverse response compensator. Then it becomes,

$$C_1(s) = \frac{bs + d}{\phi s^{\lambda+1} - as} \tag{3.39}$$

After simplifying Eq. (3.39), the controller transfer function becomes,

$$C_1(s) = \frac{1}{\phi s^{\lambda} - a}\left(b + \frac{d}{s}\right) \tag{3.40}$$

Sketch of controllers for cascade loop: Second order system in the inner loop and Integer order plus time delay (IOPDT) system in the outer loop

Inner loop controller design: Consider an inner loop plant transfer function,

$$\tilde{P}_2(s) = \frac{K_2'}{(T_2 s + 1)(T_3 s + 1)}$$

To design the inner loop controller traditional IMC controller is used. After the utilization of the Eq. (3.5) it can be written as,

$$C_2(s) = \frac{(T_3 s + 1)(T_2 s + 1)}{K_2'(\eta s + 1)^2}$$

After utilization Eq. (3.27), the expression for $P_M(s)$ becomes,

$$P_M(s) = \frac{K_1'(1 - as)}{s\left[T_1 s + 1 + K_1' T_d - asK_1' T_d\right]} \tag{3.41}$$

According to case 1 procedure, we wish to find the range of T_d.

$$\left[T_1 s^2 + s + k_1 T_d s - bs^2 T_d k_1\right] = 0 \tag{3.42}$$

All roots should be lying on the left-hand side for the system's stability. Afterward, Eq. (3.42) can be deduced as,

$$s = \frac{-(1 + K_1' T_d)}{T_1 - aK_1' T_d} < 0 \tag{3.43}$$

After simple Eq. (3.43),

$$T_d < \frac{1}{K_1'}$$

From Eq. (3.30),

$$\tilde{P}(s) = \frac{K_1'(1 - as)}{\begin{array}{l}\phi^2 s^4 (T_1 - aT_d K_1') + s^3\left(\phi^2(1 + K_1' T_d) + 2\phi(T_1 - aT_d K_1')\right)\\ +s^2 2\phi(1 + K_1' T_d + (T_1 - aT_d K_1' + s(1 + K_1' T_d)\end{array}} \tag{3.44}$$

Now converting higher-order process model into first order for simplicity, via indent toolbox in MATLAB,

$$\tilde{P}(s) = \frac{K_1'(1 - as)}{[bs + d]}$$

Further, utilizing Eq. (3.38), the controller transfer becomes,

$$C_1(s) = \frac{1}{\phi s^{\lambda} - a}\left(b + \frac{d}{s}\right)$$

TLBO algorithm

The Teaching–Learning optimization algorithm is introduced [6]. In this algorithm, exact parameters are not underlying to tune. This feature of TLBO becomes elementary. The number of generations and population size are two parameters needed in the TLBO algorithm [13]. It is segregated into the 'Teacher phase' and 'Learner phase.'

In the teaching phase

(i) For a generation of Novel outcomes (X'''_{new}) is spawned via the best solution (X'''_{best}) and mean (X'''_{mean}) of the population.
(ii) (ii) Greedy selection: Accept a new solution is better than the current one X'''.

Teaching Phase: In the teaching phase, each variable is modified as,

$$X'''_{new} = X'''_{o} + r'''_{o}(X'''_{best} - T'''_{f}X'''_{mean})$$

Here T'''_{f} is the teaching factor, which may be 1 or 2.r''' The random number can vary between 0 and 1. In the learner Phase:

(i) A new solution is generated using a partner solution X'''_{p}.
(ii) Greedy selection or randomly selected solution from population 3

$$X'''_{new} = X''' + r'''(X''' - X'''_{p})f''' < f'''_{p}$$

The TLBO algorithm flow chart is demonstrated in Fig. 3.16. The pseudocode of the TLBO algorithm is demonstrated in Appendix

$$X'_{new} = X' - r'(X' - X'_{p})f''' > f'''_{p}$$

3.6 Case Studies

Two illustrative problems are demonstrated to examine the worth of the proposed work. The simulation results are addressed for nominal and +10% process parameter variations. The efficacy of the proposed work is witnessed in terms of stability, sensitivity, and setpoint tracking. Oustaloup approximation is used with $N = 4$, $\eta_h = 10^4$, $\eta_l = 10^{-4}$ for fractional-filter implementation.

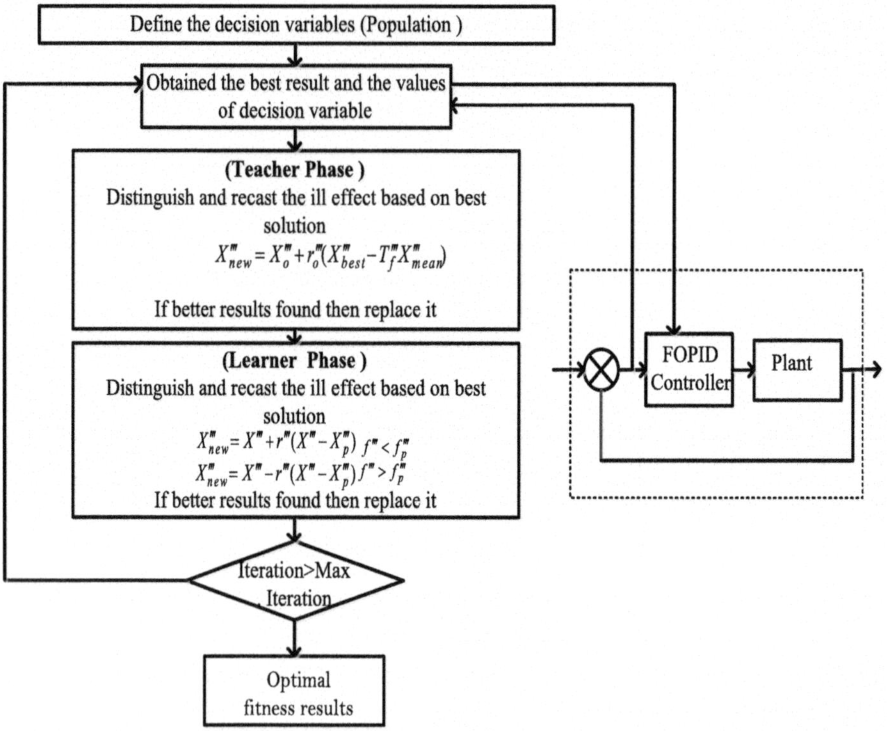

Fig. 3.16 Flow chart TLBO algorithm

3.6.1 *Case Study I*

The transfer function of the boiler drum level in a cascade framework with non-minimum phase zeros and dead–time is demonstrated [11]. However, dead-time was very small, so we have not considered it in this manuscript.

$$P_1(s) = \frac{(1 - 0.418s)}{s}, P_2(s) = \frac{0.547}{1.06s + 1}.$$

Inner loop controller design: To design the inner loop controller, consider the process.

From Eq. (3.26),

$$C_2(s) = \frac{1.06s + 1}{0.547(4s + 1)}.$$

Outer loop controller design: Initially outer loop process model is embedded with a feedback derivative controller. T_d (Eq. 3.29) is selected via Eq. (3.29). Here we assume $T_d = -0.9$. After the utilization of Eq. (3.30), it becomes,

$$P_M(s) = \frac{(1 - 0.418s)}{0.1s + 0.3762s^2}$$

From Eq. (3.33),

$$P(s) = \frac{(1 - 0.418s)}{1.5s^3 + 0.77^2 s + 0.1s}$$

Now higher-order transfer function is reduced via 'Ident Toolbox' in MATLAB. Then after utilization Eq. (3.34), it becomes,

$$P(s) = \frac{0.5(1 - 0.418s)}{1.3s + 1} \tag{3.45}$$

After utilization Eq. (3.40),

$$C_1(s) = \frac{1}{1.7s^{0.45} - 0.209}\left(1.3 + \frac{1}{s}\right) \tag{3.46}$$

Here $\tau_c = 1.7$, $\gamma = 0.45$, $\varphi = 4$, $a = 1.3$, $d = 1$, and $PM = 105$ are used.

Figure 3.17 displays a relative examination of the recommended technique with existing approaches. Figure 3.17a reveals the step response, which illustrates that the existing way suffers from peak overshoot and undershoot. You can find the transient response and performance indices in Table 3.15.

Stability analysis: The fractional characteristic polynomial for boiler drum level is given as,

$$f(s) = 1 + 1.7s^{1.45} - 0.0209s$$

The fractional characteristics polynomial has positioned the roots in $s - plane$. Here $s = \alpha + j\omega$. As discussed earlier, mapping the plane is related to the equivalent characteristic polynomial. Select $\alpha = s^{\lambda}$, The associated characteristic polynomial is

$$f(\alpha^{1/\beta}) = 1 + 1.7\alpha^{1450} - 0.0209\alpha^{1000}$$

where $s^{0.001} = \beta$.

Figure 3.18 exhibits the stability evaluation for the closed-loop system linked with the liquid-level system. Figure 3.18 analyzes that the system is stable.

Robustness Analysis: To assess the robustness of the suggested controller against process parameter variations and input disturbances, sensitivity analysis is a valuable approach. Changes in process parameters can constrain the gain margin. The parameters of the TLBO algorithm are outlined in Table 3.16.

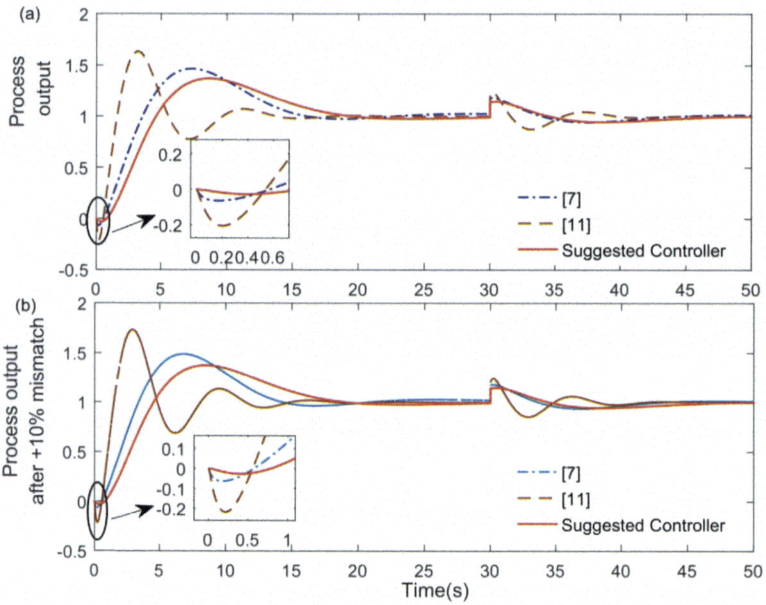

Fig. 3.17 Relative investigation for boiler drum level

Table 3.15 Transient and performance indices for boiler drum level

	Nominal case					+10% mismatch				
Method	Peak O.S. %	Peak time, s	Rise time, s	IAE	ISE	Peak O.S. %	Peak time, s	Rise time s	IAE	ISE
Suggested controller	1.37	8.1	4.44	0.01	0.0003	1.40	8.5	4.5	0.27	0.004
[7]	1.43	6.9	3.5	0.1	0.017	1.46	7.1	3.57	0.56	0.31
[11]	1.60	2.9	1.5	1.3	1.82	1.61	3.1	1.57	1.38	1.91

From Eq. (3.16)

$$S(s) = \frac{1}{1 + C_1(s)\tilde{P}(s)}$$

As a result, compute the sensitivity function substituting the value of controller and plant transfer function in Eqs. (3.33) and (3.40) into (3.16).

Figure 3.19a exhibits the absolute sensitivity analysis for the nominal case, while Fig. 3.19b shows the absolute sensitivity after the +10% mismatch case

Optimization: To determine the optimal settings for the FOPID controller, the TLBO algorithm was employed. Detailed information on the TLBO algorithm can be found in Sect. 3.3. Figure 3.20 displays the convergence curve of the objective function.

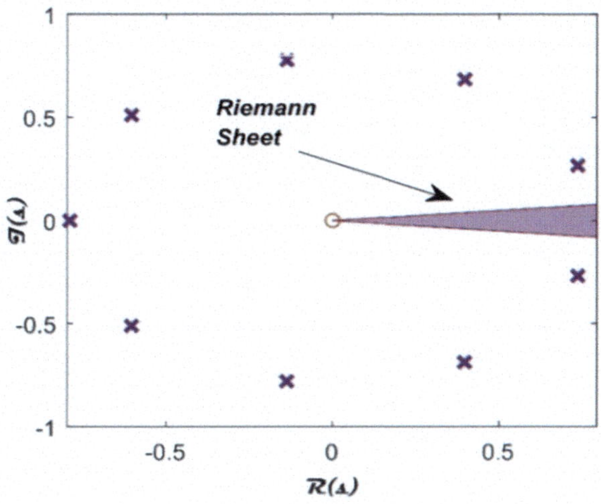

Fig. 3.18 Stability analysis of boiler drum level

Table 3.16 TLBO algorithm parameter

	No of iterations	C1	C2	Swarm size	$W_{max.}$	$W_{min.}$
Values	50	2	1.5	40	1	0.99

Figure 3.21 demonstrates that after optimization suggested controller has a better closed-loop response.

In Fig. 3.21, the FOIMC parameters are distinguished through the TLBO algorithm for controlling the boiler drum level. The MATLAB toolbox was utilized to find the optimal values for the lower and upper bounds of these parameters, which are detailed in Eq. (3.46).

$$
\begin{aligned}
1.2 &< K_p < 12 \\
0.1 &< K_i < 2 \\
0.1 &< \tau_c < 12 \\
0.05 &< \gamma < 1
\end{aligned}
$$

Afterward, an optimal value of the proposed controller settings is applied to the boiler drum level plant. Now offered optimal controller is compared with the existing controller setting.

Remarks 3.5 The offered technique is associated with approaches from references [7, 11] using a similar parameter set. The system shows inverse response behavior because of an NMP zero, necessitating an inverse response compensator. While reference [7] incorporated an inverse response compensator and a setpoint filter, leading to a trade-off between robustness and setpoint tracking, reference [11] lacked both an inverse response compensator and a fractional filter.

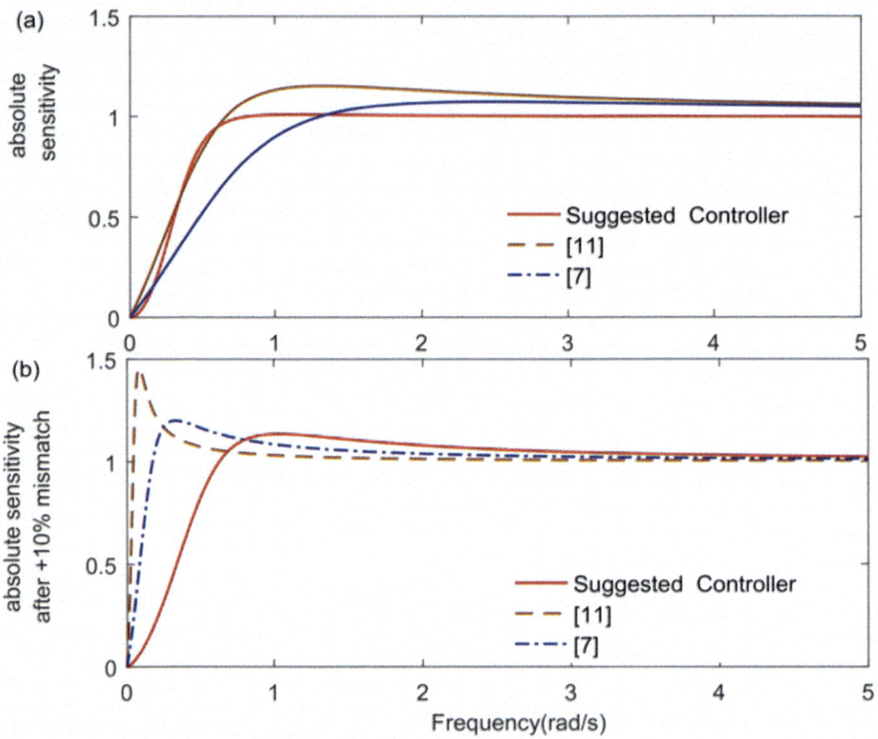

Fig. 3.19 Robustness analysis of the proposed method

Fig. 3.20 Convergence graph of the objective function

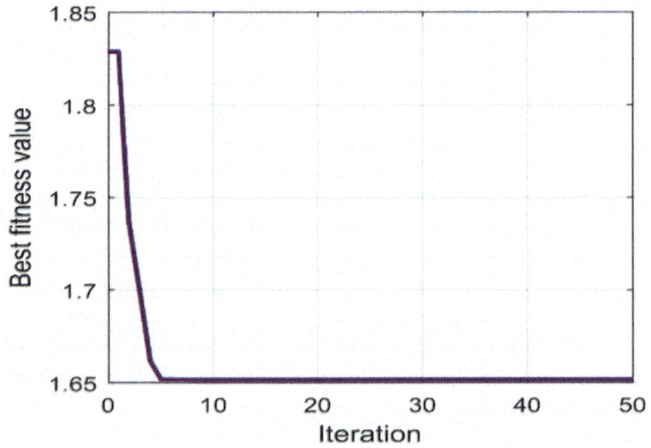

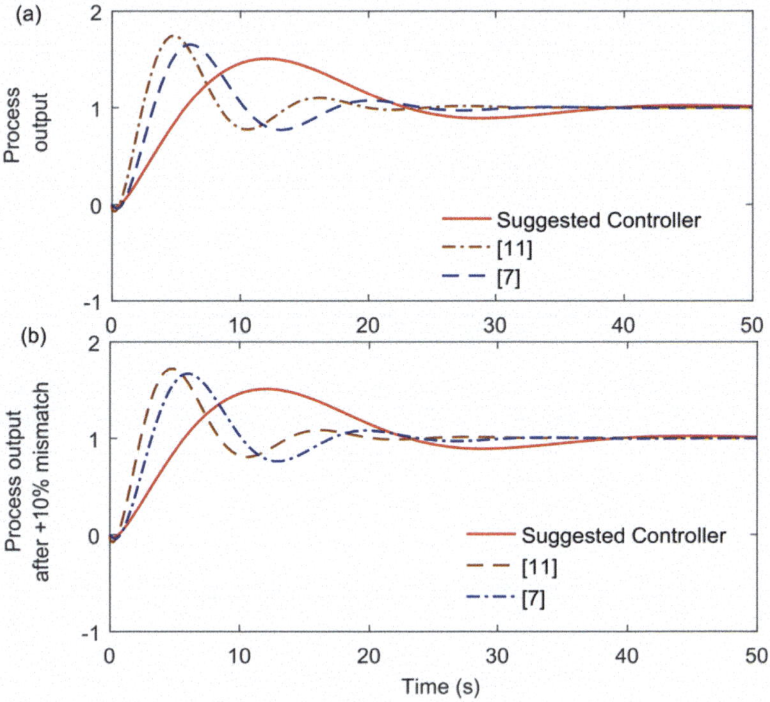

Fig. 3.21 Relative investigation after optimization

3.6.2 Case Study II

The higher-order integrating plant transfer is given [12].

$$\tilde{P}(s) = \frac{0.5(1 - 0.5s)e^{-0.7s}}{s(0.5s + 1)(0.4s + 1)(0.1s + 1)}$$

After decomposition, the plant transfer function becomes

$$\tilde{P}_1(s) = \frac{(1 - 0.5s)}{s(0.5s + 1)}, \tilde{P}_2(s) = \frac{0.5}{(0.4s + 1)(0.1s + 1)}.$$

Inner loop controller design: From Eq. (3.26), S

$$C_2(s) = \frac{(0.4s + 1)(0.1s + 1)}{0.5(3.5s + 1)}$$

Outer loop controller: Primarily outer loop process model is combined with the feedback derivative controller. T_d is selected via Eq. (3.30). Here we assume $T_d =$ 0.9. From Eq. (3.41),

$$P_M(s) = \frac{(1 - 0.5s)}{0.05s^2 + 1.9s}$$

From Eq. (3.44),

$$\tilde{P}(s) = \frac{(1 - 0.5s)}{0.45s^4 + 17s^3 + 11.45s^2 + 1.95}$$

The higher-order process transfer function was approximated using MATLAB's `ident` toolbox. After applying Eq. (3.34), the process model transfer function is derived as:

$$\tilde{P}(s) = \frac{0.65(1 - 0.5s)}{0.8s + 1} \tag{3.47}$$

From Eq. (3.40), the controller transfer function becomes

$$C_1(s) = \frac{1}{0.325s^{0.48} - 0.325}\left(0.8 + \frac{1}{s}\right)$$

The term $\frac{1}{0.325s^{0.48} - 0.325}$ shows a fractional IMC filter and $\left(0.8 + \frac{1}{s}\right)$ interprets the PI controller. Here $PM = 52$, $\lambda = 0.48$, $\phi = 0.325$, $\eta = 3.5$, $b = 0.8$, $d = 1$. Figure 3.22a illustrates step response analysis for the integrating process.

The transient response and performance measures are demonstrated in Table 3.17.

Stability analysis: An in-depth exploration of stability, utilizing the Riemann sheet principle, is discussed in Problem 3.6.1. The fractional-order characteristic polynomial for the integrating process, employed in this stability analysis, is defined as:

$$f(s) = 1 + 0.325s^{1.48} - 0.325s$$

Figure 3.23 demonstrates that all the roots of the proposed controller lie outside of the Riemann sheet, which shows the stability of the integrating process.

Robustness analysis: To assess robustness, a sensitivity analysis is performed. This requires integrating the system by substituting the values from the plant and controller transfer functions, as given in Eqs. (3.47) and (3.48), into Eq. (3.16). The robustness test for the proposed method in the context of a higher-order integrating process is depicted in Fig. 3.24. As shown in Panels Fig. 3.24a, b, the recommended controller achieves a smaller peak in both the nominal scenario and with a +10% model mismatch.

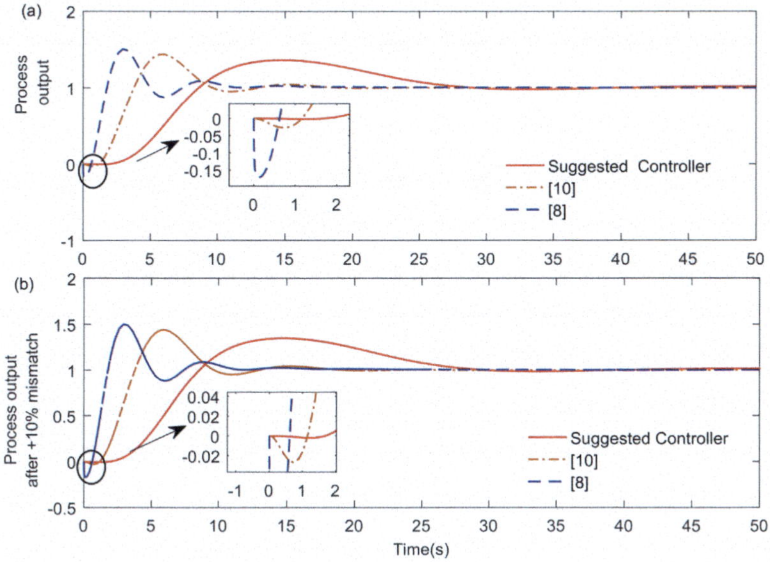

Fig. 3.22 Comparative study for higher-order integrating process

Table 3.17 Performance evaluation for integrating process

	Nominal case					+10% mismatch				
Method	Peak. o.s. %	Peak time, s	Rise time, s	IAE	ISE	Peak o.s. %	Peak time, s	Rise time s	IAE	ISE
Suggested controller	1.34	14.0	8.5	0.16	0.02	1.36	14.0	8.7	0.08	0.007
[10]	1.43	5.8	3.7	0.45	0.20	1.44	6.0	0.98	0.56	0.31
[8]	1.47	2.7	3.6	0.55	0.30	1.47	3.3	1.83	1.1	1.38

Optimization: Crucially, the optimization algorithm applied to the proposed controller utilizes the same lower and upper bounds as those in Problem 3.6.1. The objective function's convergence is depicted in Fig. 3.25.

The optimal settings for the proposed controller are then implemented on the higher-order integer plant. As shown in Fig. 3.26, this optimal controller provides better closed-loop outcomes.

Remarks 3.6 The investigation into setpoint tracking robustness, referencing works [10, 12], reveals that the proposed controller settles more quickly after disturbances. This study demonstrates that the TLBO algorithm is used to fine-tune all controller settings for equitable evaluation.

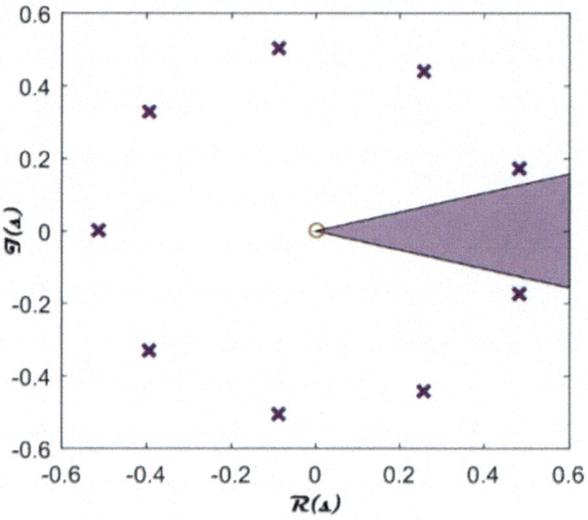

Fig. 3.23 Stability investigation for integrating higher-order process

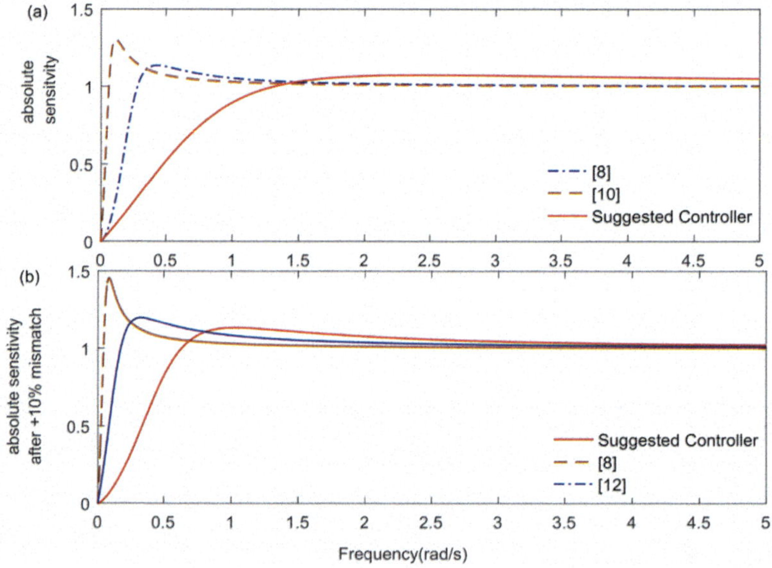

Fig. 3.24 Robustness investigation for integrating higher-order process

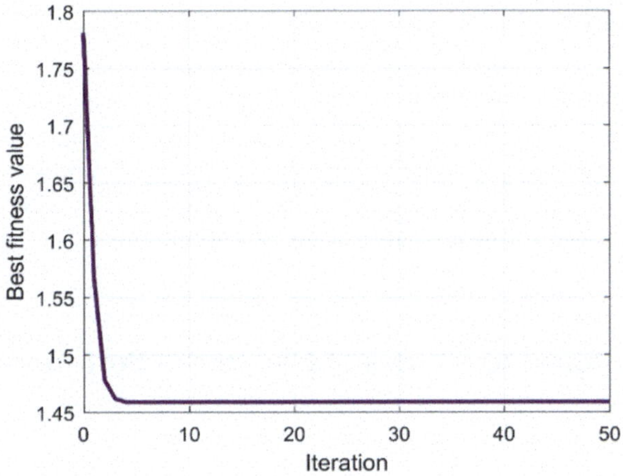

Fig. 3.25 Convergence graph of the objective function

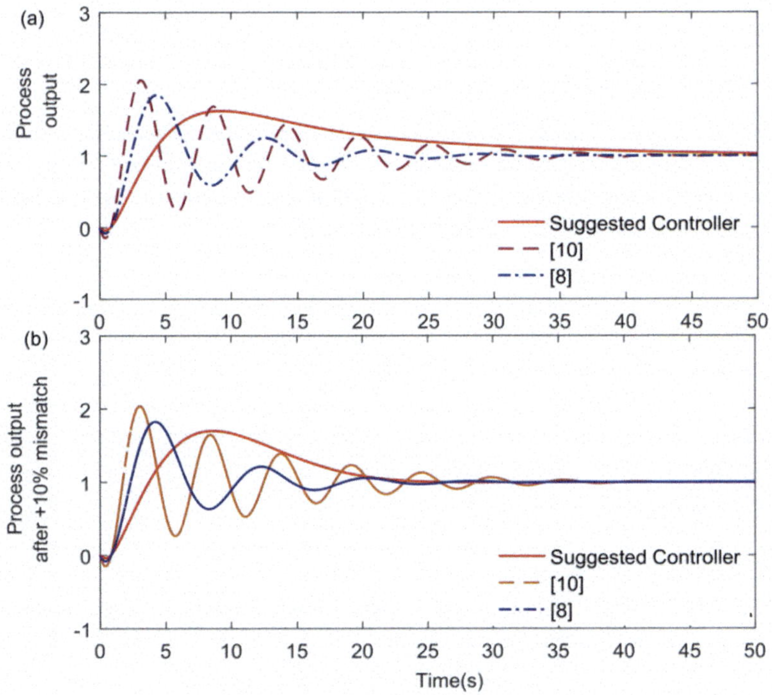

Fig. 3.26 Deviation of FOIMC constraint via TLBO algorithm

3.7 Summary

This chapter outlines a straightforward approach to enhance closed-loop responses, employing filters of different orders (higher and lower) along with higher-order approximations. However, the use of higher-order filters also presents certain limitations, which warrant further research. Chapter 4 introduces a modified parallel cascade control scheme aimed at addressing these limitations.

3.8 Some Informative Views of Chapter in Post-mortem Form

Question 1: Can you explain the concept of a series cascade control system? How does it differ from a single-loop control system, and what are its advantages and potential drawbacks?

Answer:

Concept of Series Cascade Control System

- *Definition: A series cascade control system is a control architecture where two or more controllers are arranged in a series. Typically, one controller (the primary controller) manages the overall process, while a secondary (or inner) controller manages a specific part or a subsystem of the process.*
- *Structure*
  - *Primary Controller: Controls the main process variable, providing the setpoint for the secondary controller.*
  - *Secondary Controller: Manages a more specific or secondary variable that directly influences the primary variable. The output of the secondary controller adjusts an intermediate process variable, which, in turn, affects the primary variable*

Advantages of Cascade Control Systems

- *Improved Performance*
  - *Faster Response: The secondary controller can handle disturbances and changes in the intermediate variable more quickly than the primary controller alone, improving overall system response time.*
  - *Enhanced Stability: By addressing disturbances and noise in a specific part of the process, the system can maintain stability and improve the control accuracy of the primary variable.*
- *Better Disturbance Rejection: The secondary loop can mitigate disturbances affecting the intermediate variable, leading to better overall disturbance rejection and reduced impact on the primary variable.*

- *Increased Accuracy: By controlling an intermediate variable more precisely, the primary controller can achieve a more accurate control of the main process variable.*

Potential Drawbacks

- *Increased Complexity: The design, tuning, and maintenance of a cascade control system are more complex due to the multiple controllers and interactions between them.*
- *Tuning Challenges: Properly tuning both controllers can be challenging. The secondary controller's tuning can affect the primary controller's performance, and* vice versa.
- *Potential for complication: In some cases, the added complexity might not be necessary for simpler processes where a single-loop controller could suffice. An overly complex system might introduce unnecessary complications.*

Example Application

Implementation Considerations

- *Modeling: Accurate modeling of both the primary and secondary processes is crucial for effective cascade control design.*
- *Tuning: Careful tuning of both controllers is required to ensure the cascade system performs optimally and avoids issues like oscillations or instability.*

Question 2: What are the advantages of higher-order approximations?

Answer: *The higher-order approximation can be more computationally intensive, may require more memory, and can sometimes be more sensitive to round-off errors in numerical computations. The choice between higher and lower-order approximations often depends on the specific problem at hand, balancing accuracy requirements with computational efficiency.*

Question 3: How the higher-order approximation is useful for real-world applications?

Answer: *Higher-order approximations are often used in real-world applications where more accuracy is required than what can be achieved with simpler models or first-order approximations. Here's how they are typically utilized.*

Control Systems: *In control theory, higher-order approximations are employed to design controllers that provide better performance and stability margins. For instance, in the design of PID (Proportional-Integral-Derivative) controllers, higher-order terms can be included to account for non-linearities and disturbances more accurately.*

Machine Learning: *In machine learning, higher-order approximations can improve the performance of algorithms, particularly in optimization tasks where gradient descent methods benefit from higher-order derivatives (such as Newton's method or variants like BFGS).*

Engineering and Physics: *In engineering and physics, higher-order approximations are crucial for more accurate modeling of complex systems. For example, in fluid dynamics, where the Navier–Stokes equations are solved numerically, higher-order terms can be encompassed to develop the exactness of the simulations. Similarly, in structural analysis of buildings or bridges, higher-order approximations can provide more precise predictions of stresses and deformations.*

Question 4: What are the advantages of a higher-order filter?

Answer: *Higher-order filters, such as second-order (biquadratic), third-order, or higher, offer several advantages over lower-order filters (like first-order filters) in various applications.*

1. ***Enhanced Selectivity***: *Higher-order filters can provide greater selectivity, allowing for more precise filtering of specific frequencies or bands of frequencies. This is beneficial in applications such as telecommunications, where adjacent channels need to be separated without interference.*
2. ***Compactness***: *Higher-order filters can sometimes achieve the same level of performance (such as steep roll-off or sharp transition bands) with fewer components compared to cascading multiple lower-order filters. This can result in more compact designs and lower costs in terms of components and board space.*

Question 5: What are the disadvantages of higher-order filters?

Answer: *it's imperative to note that higher-order filters also have potential disadvantages, such as increased complexity in design and implementation, higher sensitivity to quantization effects in digital implementations, and increased computational requirements. Hence, determining the appropriate filter order involves considering the particular needs of the application, striking a balance between performance, complexity, and practical constraints.*

Question 6: How would you approach solving an optimization problem in a control system or engineering context?

Answer: *Mathematical Formulation: Formulate the problem mathematically. This involves defining the objective function and constraints in terms of mathematical equations or inequalities.*

Dynamic Programming (DP): Applied to problems with a sequential decision process and where the problem can be broken down into simpler sub problems.

- *Heuristic and Metaheuristic Methods: For complex or poorly defined problems, methods like Genetic Algorithms, Simulated Annealing, or Particle Swarm Optimization may be used .*

Question 7: Can you give an example of a type of optimization problem you might encounter and the methods you would use to solve it?

- *Example Scenario: Suppose you are working on optimizing the performance of a manufacturing process. Your goal is to minimize production costs while meeting quality and production rate constraints.*
  - *Objective Function: Minimize the total cost of production, which could include costs of raw materials, labor, and overhead.*
  - *Constraints: Quality standards, production capacity limits, and available budget.*
- *Optimization Method: Depending on the complexity of the cost function and constraints, you might use:*

Implementation and Validation

- *Implementation: Apply the chosen optimization method using appropriate software tools or algorithms. Tools like MATLAB, Python (with libraries such as SciPy or PuLP), or specialized optimization software can be used.*
- *Validation: Verify the solution by checking if it meets all constraints and whether the objective function is optimized.*

Example Application in Control Systems

- *Example Scenario: In a control system, you might want to optimize the parameters of a PID controller to achieve the best balance between system stability and performance.*
  - *Objective Function: Minimize the integral of the squared error (ISE) or other performance metrics like the Integral of Absolute Error (IAE).*
  - *Constraints: Ensure the controller parameters do not lead to instability or excessive overshoot.*
  - *Method: Use techniques like optimization algorithms (e.g., gradient descent, genetic algorithms) to tune the PID parameters effectively.*

References

1. Vanavil B, Anusha AV, Perumalsamy M, Rao AS (2014) Enhanced IMC-PID controller design with lead-lag filter for unstable and integrating processes with time delay. Chem Eng Commun 201(11):1468–1496
2. Lino P, Maione G (2018) Cascade fractional-order pi control of a linear positioning system. IFAC-OnLine 51(4):557–562
3. Chekari T, Mansouri R, Bettayeb M (2018) IMC-PID fractional order filter multi-loop controller design for multivariable systems based on two degrees of freedom control scheme. Int J Control Autom Syst 16(2):689–701

4. Ranganayakulu R, Rao AS, Bhasakar B (2020) Analytical design of fractional IMC filter—PID control strategy for performance enhancement of cascade control systems. Int J Syst Sci 51(10):1699–1713
5. Yadav M, Patel HG, Nagarsheth SH (2024) Optimal fractional-order series cascade controller design: a refined Bode's ideal transfer function perspective. Trans Inst Meas Control 46(10):1955–1965
6. Singhal AK, Yadav M, Yadav V, Deshmukh J, Billore M, Mazumdar H (2024) Optimizing the series cascade control structure for nonminimum phase system regulation. Appl Res e202300051
7. Nagarsheth SH, Sharma SN (2020) Control of non-minimum phase system with dead time: a fractional system viewpoint. Int J Syst Sci 51(11):1905–1928
8. Begum KG, Rao AS, Radhakrishnala TK (2017) Enhanced IMC-based PID controller design for non-minimum phase (NMP) integrating processes with time delays. ISA Trans 68:223–234
9. Stephanopoulos G (1984) Chemical process control. Englewood Cliffs, New Jersey, Prentice Hall
10. Pai NS, Chang SC, Huang CT (2010) Tuning PI/PID controllers for integrating processes with dead-time and inverse response by simple calculations. J Process Control 20(6):726–733
11. Rao AS, Seethaladevi S, Uma S, Chidambaram M (2009) Enhancing the performance of parallel cascade control using smith predictor. ISA Trans 48:220–227
12. Joshi D, Deb D, Giri AK (2024) MRAS disturbance observer-based sensorless field-oriented backstepping control of BLDC motor drive. In: Electrical engineering. Springer, Berlin. https://doi.org/10.1007/s00202-024-02378
13. Kundu S, Singh M, Giri AK (2022) Implementation of in-loop filter based DSC-FLL control approach for power quality improvement in isolated microgrid. In: Proceedings of 2022 IEEE 31st international symposium on industrial electronics (ISIE). USA. https://doi.org/10.1109/ISIE51582.2022.9831712
14. Giri AK, Arya SR, Maurya R, Hybrid order generalized integrator based control for VSC to improve the PMSG operation in isolated mode. In: The proceedings of 2020 first international conference on power, control and computing technologies (ICPC2T). India. https://doi.org/10.1109/ICPC2T48082.2020.9071498
15. Kundu S, Singh M, Giri AK (2024) Synchronization and control of WECS-SPV-BSS-based distributed generation system using ICCF-PLL control approach. Electric Power Syst Res 226:109919
16. Giri AK, Arya SR, Guerrero JM, Kumar S, Power quality: infrastructures and control. Springer Nature, Singapore. ISBN: 978-981-19-7956-9
17. Giri AK, Arya SR, Vinnikov D, Distributed energy systems design, modeling, and control, vol 1. CRC Press, pp 1–308. ISBN: 9781032134246
18. Arya SR, Patel MM, Alam SJ, Srikakolapu J, Giri AK, Chitti Babu B (2021) Classical control algorithms for permanent magnet synchronous generator driven by diesel engine for power quality. Int J Circuit Theory Appl 49(3):576–601
19. Kundu S, Giri AK, Singh M, Kadiyan S, Adaptive control strategy for isolated renewable energy-based generation system with intermittency. In: Electrical engineering. Springer, Belin, https://doi.org/10.1007/s00202-024-02454-0
20. Meena DC, Singh M, Giri AK (2021) Leaky-momentum control algorithm for voltage and frequency control of three-phase SEIG feeding isolated load. J Eng Res ICARI Spec 109:120
21. Meena DC, Singh M, Giri AK (2021) A modified NLMS control algorithm for coordinated operation in three-phase wind-energy conversion system. Int J Renew Energy Res (IJRER) 11(4):1621–1629
22. Dalwadi N, Deb D, Ozana S (2023) Performance evaluation of RW-quadrotor and Bi-quadrotor for payload delivery. IET Intel Transport Syst 17(11):2221–2236
23. Dalwadi N, Deb D, Ozana S, Adaptive hybrid control of quadrotor drones, vol 461, pp 1–174. ISBN: 978-981-19-9743-3

Chapter 4
Improved Parallel Cascade Control for Non-minimum Phase System

4.1 An Enhancement in Parallel Cascade Scheme

This chapter emphasizes the development of an enhanced parallel cascade control framework tailored for a non-minimum phase system. Within a parallel cascade configuration, disturbances and the manipulated variable can sequentially impact both loops [1].

Some process control systems demonstrate an initial response that opposes the zero-frequency gain. This inverse response phenomenon often occurs due to an odd multiplicity of Non-Minimum Phase (NMP) open-loop zeros [2]. Additionally, [3] proposed an advanced parallel cascade control method incorporating a Smith predictor across various case studies. Lastly, [4] discussed fractional-order controllers equipped with dead-time compensators tailored for parallel cascade control strategies.

The studies mentioned propose outer loop controllers for parallel cascade configurations using fractional filters and dead-time compensators, as documented in the literature. However, there is currently no unified theory presented for an outer loop controller integrating higher-order fractional filters, compensation for inverse response, and dead-time compensators in parallel cascade control systems.

4.2 Parallel Cascade Scheme

Figure 4.1 illustrates a parallel cascade control configuration, where both the manipulated variable and disturbances affect the responses of both the outer and inner loops simultaneously. The output of the inner loop controller resets the setpoint for the inner loop, influencing two variables.

In parallel cascade control, the dynamics of the inner loop must be significantly faster than those of the outer loop. This ensures that disturbances entering the outer

M. Yadav et al., *Control Strategies for Non-minimum Phase Systems*,
Studies in Infrastructure and Control, https://doi.org/10.1007/978-981-96-2351-8_4

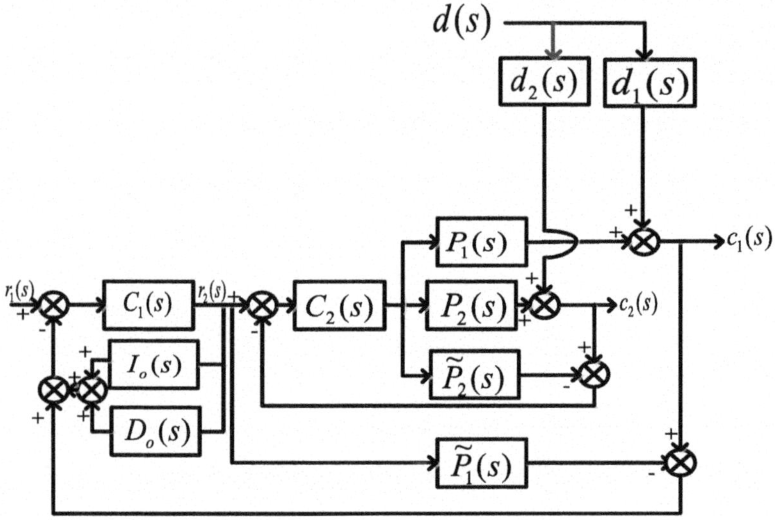

Fig. 4.1 Proposed cascade control structure

loop are promptly attenuated, minimizing steady-state errors. Therefore, the inner loop controller is typically tuned first, followed by tuning the outer loop controller in conjunction with the inner loop controller [2].

In Fig. 4.1, the dead-time compensator $D_o(s) = \frac{1-e^{-t_d s}}{e^{-t_d s}}\tilde{P}_1(s)$ and $I_o(s) = 2as\tilde{P}_1(s)$ inverse response compensator are combined with a feedback controller. Here t_d, a, and $\tilde{P}_1(s)$ are the dead-time, NMP zero, and the transfer function of the outer loop, respectively. Since both the loops are decoupled entirely for setpoint and load disturbances, it requires two controllers, i.e., $C_1(s)$ and $C_2(s)$. In the process industries, the inner loop dynamics is typically a stable environment and the outer loop dynamics is non-minimum phase-stable/unstable/integrating (Stephanopoulos 2015). According to Fig. 4.1, the transfer function of closed-loop response can be determined as,

$$\frac{c_2(s)}{r_2(s)} = C_2(s)P_2(s), \tag{4.1}$$

$$\frac{c_2(s)}{d_2(s)} = (1 - P_2(s)C_2(s))d_2(s), \tag{4.2}$$

$$\frac{c_1(s)}{r_1(s)} = \frac{d_1(s)\begin{pmatrix} 1 + C_2(s)P_2(s) - C_2(s)\tilde{P}_2(s) \\ +C_1(s)(I_o(s) + D_o(s)) \end{pmatrix}}{\begin{matrix} 1 + C_1(s)(I_o(s) + D_o(s)) - C_1(s)\tilde{P}_1(s) \\ +C_1(s)C_2(s)P_1(s) + C_2(s)P_2(s) - C_2(s)\tilde{P}_2(s) \end{matrix}}, \tag{4.3}$$

4.3 Controller Design

In this work, inverse response and dead-time compensators are also embedded with the outer loop controller of a parallel cascade scheme. The generalized controller with a fractional filter coupled with dead time, an inverse response compensator in the IMC framework [2],

$$C_1(s) = \frac{\tilde{P}^+(s)e^{-t_d s}}{\tilde{P}(s)e^{-t_d s}\{(1+\phi s^{\lambda+1})^n - I_o(s) - \tilde{P}^+(s)\} - \tilde{P}^+(s)\tilde{P}(s)(1-e^{-t_d s})}. \quad (4.4)$$

Here $\tilde{P}^+(s)$, t_d are the system transfer function's non-inverted part and time delay terms.

Design of controllers for cascade loop: First order plus time delay (FOPTD) System in the inner loop and Integer order plus time delay (IOPDT) system in the outer loop

$$\tilde{P}_1(s) = \frac{K_1'(1-as)}{s}e^{-t_{d1}s}, \tilde{P}_s(s) = \frac{K_2'e^{-t_{d2}s}}{s+1}.$$

Here K_1', K_2' is the gain of the plant transfer function.

Inner loop controller design: Here, the process model of the inner loop is decomposed according to the IMC scheme [2],

$$\tilde{P}_2(s) = \tilde{P}_2^-(s)\tilde{P}_2^+(s).$$

A transfer function consists of the minimum phase ($\tilde{P}_2^-(s)$) characteristic of the process model, i.e., $\frac{K_2'}{s+1}$, while $\tilde{P}_2^+(s)$ is a non-invertible feature of the process model. In the IMC procedure, the inverted portion is utilized to design the IMC controller, which can be written as [2],

$$C_2(s) = \frac{f_1(s)}{\tilde{P}_2^-(s)}. \quad (4.5)$$

Here $f_1(s)$ indicates a low pass filter, i.e., $f_1(s) = \frac{1}{(1+\eta s)^n}$. Here n can be utilized to make the controller proper and η is the tuning parameter, which tunes based on maximum sensitivity M_s. [2],

Outer loop controller design:

In parallel cascade control, the output of the inner loop controller adjusts the inner loop's setpoint, and this controller's output impacts two output variables. As a result, the overall transfer function of the plant is modified.

$$\tilde{P}(s) = C_2(s)\tilde{P}_1(s). \quad (4.6)$$

According to the IMC procedure, the outer loop process model is decomposed into the invertible $\tilde{P}^{-}(s) = \frac{K(s+1)}{s(\eta s+1)}$ and non-invertible parts $\tilde{P}^{+}(s) = (1 - as)e^{-t_{d1}s}$. The delay term $e^{-t_{d1}s}$ can be approximated via second-order Pade approximation. The strategy for the inverse. Here $K = \frac{K_1}{K_2}$.

$$C_{f_1}(s) = \frac{(\eta s+1)\left(1 + \frac{t_d s}{2} + \frac{t_d^2 s^2}{12}\right)}{K\left(\begin{array}{l} \frac{t_d^2 a s^4}{12} + s^3\left(\left(\frac{t_{d_1}a}{2} - \frac{t_{d_1}}{12}\right) + \frac{t_d^2 a}{12}\right) \\ +s^2\left(\left(\frac{t_{d_1}a}{2} - \frac{t_d^2}{12}\right) - \left(\frac{t_{d_1}}{2} + a\right)\right) + \phi s^{\lambda+2} + \phi s^{\lambda+1} - s\left(\frac{t_{d_1}}{2} + a\right) \end{array}\right)}. \tag{4.7}$$

After utilizing the second-order fractional IMC filter $f_2'(s) = \frac{\lambda s+1}{(\alpha s^{\delta+1}+1)^2}$, Pade's approximation of the controller transfer function becomes,

$$C_{f_2'}(s) = \frac{s(\eta s+1)(\tau s+1)\left(1 + \frac{t_d s}{2} + \frac{t_d^2 s^2}{12}\right)}{K\left(\begin{array}{l} \frac{t_d^2 a\tau s^5}{12} + s^4\left(\left(\frac{t_d\tau a}{2} - \frac{t_d^2\tau}{2} + \frac{t_d^2 a}{12}\right) + \frac{t_d^2}{12}\right) + s^3\left(\begin{array}{l}\left(\frac{t_d a}{2} - \tau a - \frac{t_d\tau}{2} - \frac{t_d^2}{12}\right) \\ +\left(\frac{t_d\lambda b}{2} - \frac{t_d^2\lambda}{12} - \frac{t_d^2}{12}\right)\end{array}\right) + \\ s^2\left(\begin{array}{l}\left(\frac{t_d a}{2} - \tau a - \frac{t_d\tau}{2} - \frac{t_d^2}{12}\right) \\ -(\tau + a)\end{array}\right) \\ +\phi^2 s^{2\lambda+3} + 2\phi s^{\lambda+2} + \phi^2 s^{2\lambda+2} + 2\phi s^{\lambda+1} - s(\frac{t_d}{2} + a) \end{array}\right)}. \tag{4.8}$$

Here τ is the additional degree of freedom [2]

$$\lambda = \alpha\left[1 - \sqrt{\left(1 - \frac{\delta}{\alpha}\right)^3 e^{-\frac{\gamma}{\alpha}}}\right]. \tag{4.9}$$

Fractional filter parameters ϕ, λ can be tuned via Bode's ideal transfer function. Bode's ideal transfer function uses the phase margin (P.M.) and gains cross-over frequency. It can be termed [3],

$$\lambda = \frac{\pi - PM}{\frac{\pi}{2}} - 1, \phi = \frac{1}{(\omega)^{\lambda+1}} \tag{4.10}$$

Design of controllers for cascade loop: First order plus time delay (FOPTD) system in the inner loop and the FOPDT system in the outer loop

$$\tilde{P}_1(s) = \frac{K_1(1 - as)}{T_1 s + 1}e^{-t_{d1}s}, \tilde{P}_2(s) = \frac{K_2'}{T_2 s + 1}e^{-t_{d_2}s}.$$

Inner loop controller design: The inner loop controller is designed via Eq. (4.5). It can be written as,

$$C_2(s) = \frac{T_2 s + 1}{K_2'(\eta s + 1)}.$$

Outer loop controller design: The overall plant transfer function of the outer loop a process model is achieved after using Eq. (4.7),

$$P(s) = \frac{K_1'(T_2 s + 1)(1 - as)}{K_2'(\eta s + 1)(T_1 s + 1)} e^{-t_{d_1} s}. \tag{4.11}$$

Now utilize the first-order fractional IMC filter $f_2(s) = \frac{1}{(1+\phi s^{\lambda+1})^n}$ for the outer loop controller design of parallel cascade control. After using Eq. (4.8), the controller transfer function for the outer loop process model becomes,

$$C_{f_2}(s) = \frac{(\eta s + 1)(T_1 s + 1)\left(1 + \frac{t_d s}{2} + \frac{t_d^2 s^2}{12}\right)}{K\begin{pmatrix}\frac{T_2 t_d^2 a s^4}{12} + s^3\left(\frac{T_2 a t_d}{2} + \frac{a t_d^2}{12} - \frac{T_2 t_d^2}{12}\right) + \\ s^2\left(\left(\frac{t_d a}{2} - T_2 a_1 - \frac{t_d^2}{12}\right)\right)\end{pmatrix}}. \tag{4.12}$$

After utilizing the second-order fractional IMC filter $f_2(s) = \frac{\tau s+1}{(\phi s^{\lambda+1}+1)^2}$ and Pade approximation, the controller transfer function becomes (Table 4.1).

$$C'_{f_2}(s) = \frac{(\tau s + 1)(\eta s + 1)(T_1 s + 1)\left(1 + \frac{t_d s}{2} + \frac{t_d^2 s^2}{12}\right)}{K\begin{pmatrix}\frac{T_2 t_d^2 a \tau s^5}{12} + s^4\left(T_2\left(\frac{t_d \tau a}{2} - \frac{t_d^2}{12} + \frac{t_d^2 \tau}{12}\right) + \frac{t_d^2 a \tau}{12}\right) + \\ s^3\begin{pmatrix} T_2\left(\frac{t_d a}{2} - \tau a - \frac{t_d \tau}{2} - \frac{t_d^2}{12}\right) \\ + \left(\frac{t_d \tau a}{2} + \frac{t_d^2 \tau}{12} - \frac{t_d^2}{12}\right)\end{pmatrix} + s^2\begin{pmatrix}\left(\frac{t_d a}{2} - a\tau - \frac{t_d \tau}{2} - \frac{t_d^2}{12}\right) \\ -T_2\left(\tau + a + \frac{t_d}{2}\right)\end{pmatrix} + \\ T_2 \phi^2 s^{2\lambda+3} + 2\phi T_2 s^{\lambda+2} + \phi^2 s^{2\lambda+2} + 2\phi s^{\lambda+1} - s\left(\frac{t_d}{2} + a + \tau\right)\end{pmatrix}} \tag{4.13}$$

4.4 Case Studies

Two systems are utilized to evaluate the performance of the suggested approach: (i) Boiler drum level system and (ii) Distillation column.

Table 4.1 Filters and approximations for the parallel cascade control

Method	Fractional filter + PID controller
1st order filter	$f_1(s) = \frac{1}{\eta s+1}$ (inner loop), $f_2(s) = \frac{1}{(\phi s^{\lambda+1}+1)}$ (outer loop) + 2nd order Pade approximation
2nd order filter	$f_1(s) = \frac{1}{\eta s+1}$ (inner loop), $f_2(s) = \frac{\tau s+1}{(\phi s^{\lambda+1}+1)^2}$ (outer loop) + 2nd order Pade approximation

4.4.1 Case Study I

The regulation of drum level involves controlling the influence of boiler feed water. According to [2], the transfer function describing the boiler drum level system can be formulated as follows:

$$\tilde{P}_1(s) = \frac{(1-0.418s)e^{-0.07s}}{s}, \tilde{P}_2(s) = \frac{0.547e^{-0.03s}}{1.06s+1}.$$

Inner loop controller design: For the inner loop, controller settings utilize Eq. (4.5). It can be written as,

$$C_2(s) = \frac{1.06s+1}{0.547(0.5s+1)}.$$

Outer loop controller design: After the utilization of Eq. (4.6), the overall plant transfer function can be obtained as,

$$\tilde{P}(s) = \frac{1.8(1-0.418s)(1+1.06s)e^{-0.07s}}{s(0.5s+1)}. \tag{4.14}$$

The suggested controller settings for the outer loop in the fractional IMC framework can be obtained via Eq. (4.7),

$$C_{f_1}(s) = \frac{0.5s^2+s}{1.8(0.00018s^3+0.013s^2+2.6s^{1.45}-0.46s+2.5s^{0.45}-0.453)}\left(0.35+\frac{0.099}{s}+0.00035s\right). \tag{4.15}$$

where the term $\frac{0.5s^2+s}{1.8(0.00018s^3+0.013s^2+2.6s^{1.45}-0.46s+2.5s^{0.45}-0.453)}$ shows fractional filter, and the term $\left(0.35+\frac{0.099}{s}+0.00035s\right)$ exhibits PID controller with $PM = 49.2$, $\lambda = 0.45$, $\tau = 0.5$ and $\phi = 2.5$. The suggested other controller settings for the outer loop in the fractional IMC framework with a second-order fractional IMC filter can be obtained via Eq. (4.8).

$$C_{f_2}(s) = \frac{0.375s^3 + 1.25s^2 + s}{1.8(0.00014s^4 + 0.00924s^3 - 0.36s^2 - 0.65s - 0.45 + 2.65s^{3.90} + 5.3s^{2.45} + 6.25s^{2.90} + 5s^{1.45})} \left(0.35 + \frac{0.099}{s} + 0.00035s\right). \tag{4.16}$$

Figure 4.2 depicts a comparative analysis of 1st and 2nd-order fractional filters using a second-order Pade approximation for designing the outer loop controller within the IMC framework.

Table 4.2 indicates that the proposed controller achieves lower values of IAE and ISE.

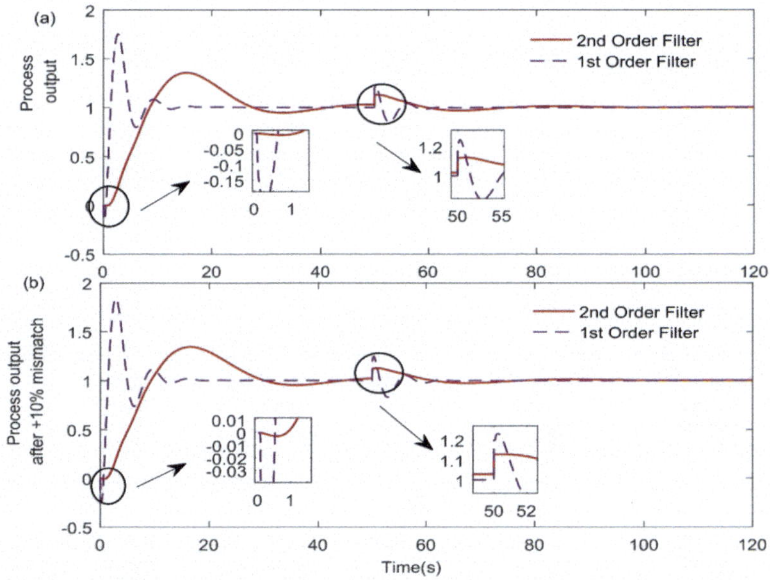

Fig. 4.2 Step response for liquid level system

Table 4.2 Performance indices for the liquid level system

Controller	Nominal case		+10% Perturbed case	
	IAE	ISE	IAE	ISE
Proposed controller	0.16	0.028	0.23	0.056
Uma et al. [1]	0.2	0.04	0.3	0 0.09
Nagarsheth and Sharma [2]	0.3	0010 0.10	0.38	0.14
Raja and Ali [3]	0.4	0.16	0.43	0.18

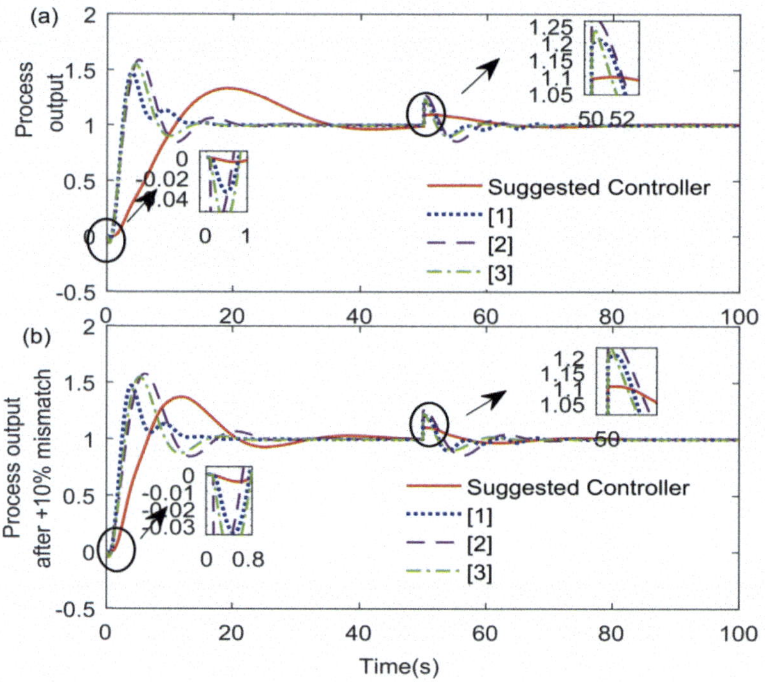

Fig. 4.3 Comparative analysis for liquid level system

Figure 4.2a shows the step response comparison between the 1st and 2nd-order filters. Figure 4.2b illustrates the closed-loop response under a +10% variation in process parameters

Figure 4.3 illustrates the step response of the cascade system with a disturbance magnitude of 0.2 at t = 50 s, depicted for both nominal and +10% mismatch scenarios.

Additionally, the stability analysis of the proposed scheme relies on the Riemann surface, which is graphically evaluated using the FOMCON toolbox in MATLAB. The general expression for stability analysis of the first-order fractional IMC filter can be formulated as follows [2]:

$$f_2(s) = 1 + \frac{t_{d1}^2 a s^4}{12} + s^3\left(\left(\frac{t_{d_1}a}{2} - \frac{t_{d1}^2}{12}\right) + \frac{t_{d1}^2 a}{12}\right) + s^2\left(\left(\frac{t_{d_1}a}{2} - \frac{t_d^2}{12}\right) - \left(\frac{t_{d_1}}{2} + a\right)\right) + \phi s^{\lambda+2} + \phi s^{\lambda+1} - s\left(\frac{t_{d_1}}{2} + a\right).$$

Likewise, the generalized expression for stability analysis with 2nd order fractional IMC the filter is given as,

$$
\begin{aligned}
f_2'(s) &= 1 + \frac{t_{d1}^2 a \tau s^5}{12} + s^4\left(\left(\frac{t_{d_1}\tau a}{2} - \frac{t_{d1}^2}{12} + \frac{t_d^2 a}{12}\right) + \frac{t_{d1}^2 a \tau}{12}\right) \\
&+ s^3\left(\left(\frac{t_{d_1} a}{2} - \tau a - \frac{t_{d_1}\tau}{2} - \frac{t_{d1}^2}{12}\right) + \left(\frac{t_{d_1}\tau a}{2} + \frac{t_d^2 \tau}{12} - \frac{t_d^2}{12}\right)\right) \\
&+ s^2\left(\left(\frac{t_{d_1} a}{2} - \tau a - \frac{t_{d_1}\tau}{2} - \frac{t_{d1}^2}{12}\right) - \left(\tau + a + \frac{t_{d_1}}{2}\right)\right) \\
&+ \phi^2 s^{2\lambda+3} + 2\phi s^{\lambda+2} + \phi^2 s^{2\lambda+2} + 2\phi s^{\lambda+1} - s\left(\frac{t_{d1}}{2} + a + \tau\right).
\end{aligned}
$$

The specific structure of fractional characteristic polynomial for boiler drum level system can be written as,

$$
f_2(s) = 1 + 0.00018s^4 + 0.013s^3 - 0.46s^2 - 0.453s + 2.65s^{2.45} + 2.5s^{1.45}.
$$

$$
\begin{aligned}
f_2'(s) &= 1 + 0.00014s^5 + 0.00924s^4 - 0.36s^3 - 0.65s^2 - 0.45s \\
&+ 2.65s^{4.90} + 5.3s^{3.45} + 6.25s^{3.90} + 5s^{2.45}.
\end{aligned}
$$

The roots of fractional-order characteristic polynomials lie in the $s = \alpha + j\omega$. The mapping of $s - plane$ a plane linked with an equivalent characteristic polynomial. Choose $\alpha = s^\beta$, then linked characteristic polynomial becomes,

$$
\begin{aligned}
f_2(\alpha^{1/\beta}) &= 1 + 0.00018\alpha^{4000} + 0.013\alpha^{3000} - 0.46\alpha^{2000} \\
&- 0.453\alpha^{1000} + 2.65\alpha^{2450} + 2.5\alpha^{1450}.
\end{aligned}
$$

$$
\begin{aligned}
f_2'(\alpha^{1/\beta}) &= 1 + 0.00014\alpha^{5000} + 0.00924\alpha^{4000} - 0.36\alpha^{3000} - 0.65\alpha^{2000} \\
&- 0.45\alpha^{1000} + 2.65\alpha^{4900} + 5.3\alpha^{3450} + 6.25\alpha^{3900} + 5\alpha^{2450}.
\end{aligned}
$$

Where $s^{0.001} = \alpha$. From the stability point of view of the recommended fractional closed-loop system, the absolute values of the angles of roots of associated characteristics polynomial should be more significant $\frac{\beta\pi}{2}$.

The graphical analysis is depicted in Fig. 4.4. Since all the roots lie outside the Riemann sheet, it confirms the stability of the system [2].

The sensitivity analysis confirms the attenuation of disturbances occurring at the sensitivity function $|s(j\omega)| \leq 1$ and the amplification of disturbances occurring $|s(j\omega)| \geq 1$. The sensitivity function is given by [3],

$$
S(s) = \frac{1}{1 + \tilde{P}(s)C(s)}. \tag{4.17}
$$

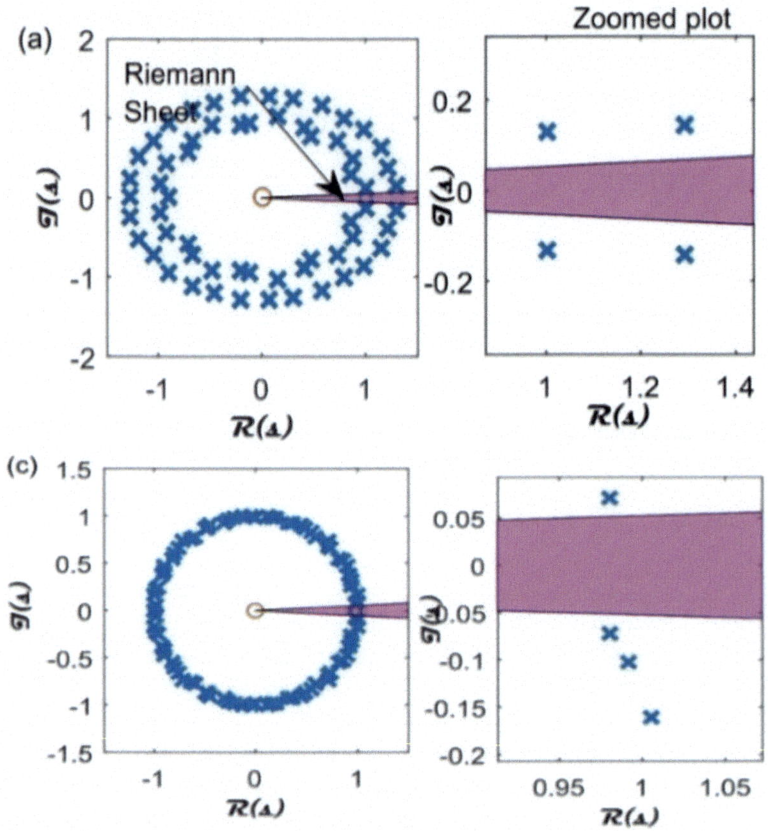

Fig. 4.4 Stability analysis of the liquid level system

Figure 4.5 illustrates a graphical representation of the absolute sensitivity. The maximum absolute sensitivity serves as a reliable measure of robustness, where a lower value indicates greater robustness.

Remark 4.1 To assess the effectiveness of the proposed controller, it is compared with the approaches in [1–3]. Their methods employ lower-order filters and approximations for controller design. Specifically, [1] utilizes a lower-order filter along with a setpoint filter. In contrast, this study employs a higher-order fractional IMC filter, eliminating the need for a setpoint filter. In [2, 3], inverse response and dead-time compensators were not included in the controller design.

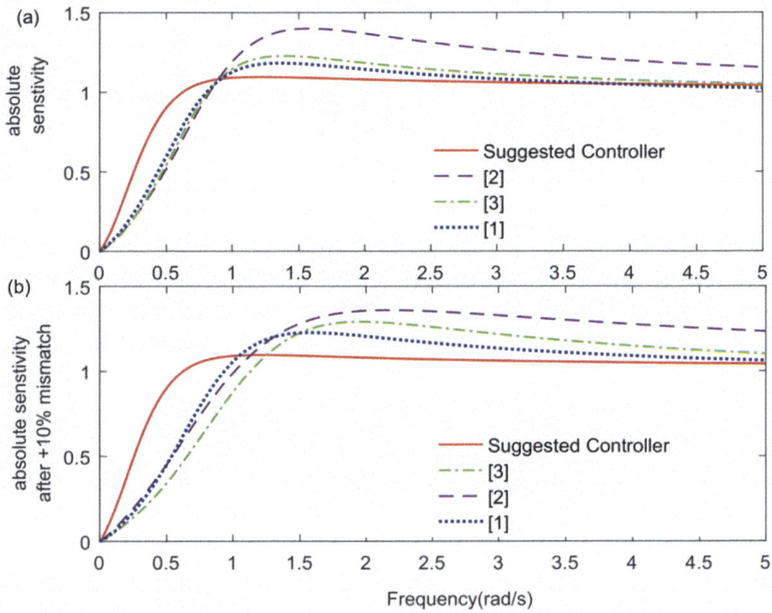

Fig. 4.5 Robustness analysis for the liquid level system

4.4.2 *Case Study II*

The transfer function describing the distillation column is detailed in [3].

$$\tilde{P}_1(s) = \frac{(1-0.2s)e^{-0.1s}}{s+1}, \tilde{P}_2(s) = \frac{e^{-0.1s}}{s+1}.$$

Inner loop controller design: The inner controller is obtained from Eq. (4.5). It can be written as,

$$C_2(s) = \frac{s+1}{(0.3s+1)}.$$

After the utilization of Eq. (4.11), the overall plant transfer function is modified as,

$$\tilde{P}(s) = \frac{(1-0.2s)(1+s)e^{-0.2s}}{(s+1)(0.3s+1)}. \tag{4.18}$$

Outer loop controller design: The suggested controller settings for the outer loop in the fractional IMC framework can be obtained via Eq. (4.12).

It can be written as,

$$C_{f_2}(s) = \frac{0.3s^2 + 1.3s + 1}{(0.0033s^3 + 0.099s^2 - 0.29s - 0.3 + 0.65s^{1.27} + 0.65s^{0.27})}\left(0.05 + \frac{1}{s} + 0.00083s\right). \quad (4.19)$$

where the term $\frac{0.3s^2+1.3s+1}{(0.0033s^3+0.099s^2-0.29s-0.3+0.65s^{1.27}+0.65s^{0.27})}$ exhibits a fractional filter $\lambda = 0.27$, $\tau = 0.3$ and $\phi = 0.65$. The term $\left(0.05 + \frac{1}{s} + 0.00083s\right)$ displayed display PID controller. The suggested other controller settings for the outer loop in fractional IMC framework with second-order fractional IMC filter can be obtained via Eq. (4.13).

$$C_{f_2}(s) = \frac{0.12s^3 + 0.82s^2 + 1.7s + 1}{(0.00007s^4 + .00337s^3 - 0.04s^2 - 0.70s - 0.55 + 0.42s^{2.54} + 1.30s^{1.27} + 0.42s^{1.54} + 1.30s^{0.27})}\left(0.05 + \frac{1}{s} + 0.00083s\right) \quad (4.20)$$

Figure 4.6 displays a relative analysis of 1st and 2nd-order fractional filters with second-order Pade approximation for the outer loop in the IMC framework.

In Fig. 4.7a, b, it is evident that the proposed controller exhibits quicker settling times following disturbances. Figure 4.7 illustrates that the settling performance of the proposed controller yields significant improvements in outcomes.

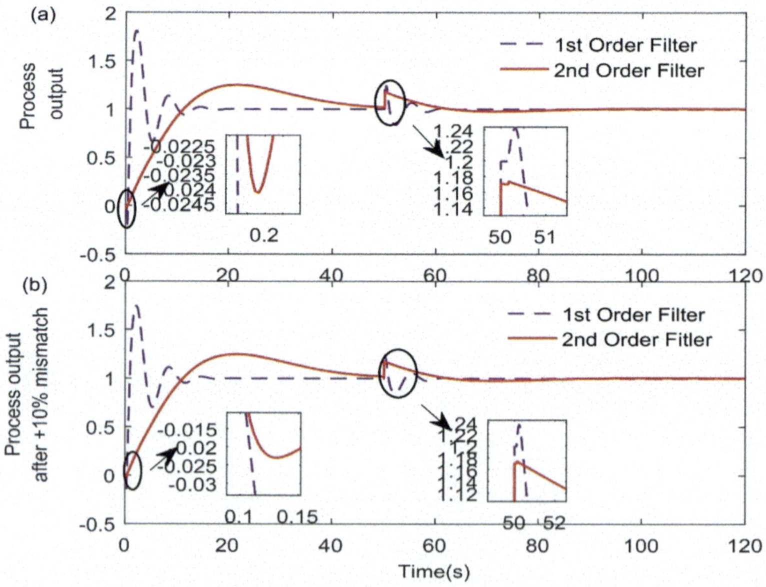

Fig. 4.6 Closed-loop response for the distillation column

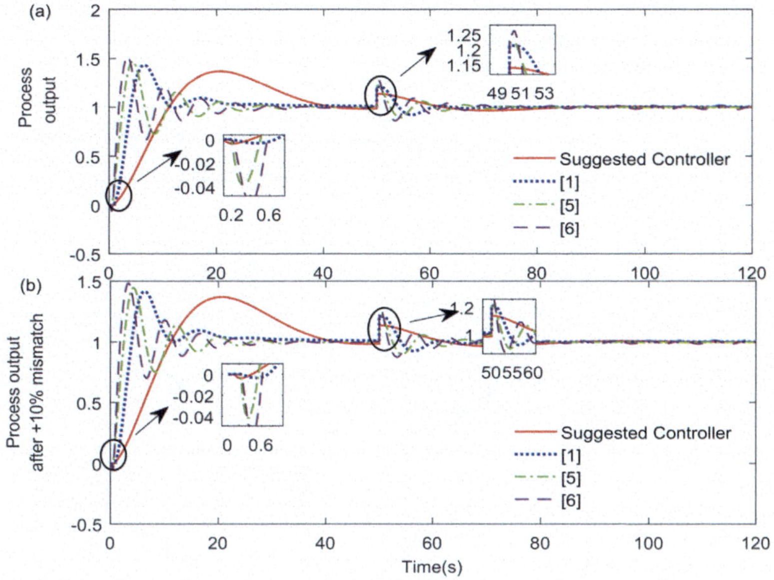

Fig. 4.7 Comparative study for the distillation column

The description of the fractional characteristic polynomial for the distillation column after utilizing 1st order fractional IMC filter and 2nd-order Pade approximation can be written as,

Likewise, fractional characteristic polynomial for the distillation column after utilizing 2nd order fractional IMC filter and 2nd order Pade approximation can be written as,

$$f_2(s) = 1 + 0.00007s^5 + .00337s^4 - 0.04s^3 - 0.70s^2 - 0.55s + 0.42s^{3.54}$$
$$+1.30s^{2.27} + 0.42s^{2.54} + 1.30s^{1.27}.$$

The detailed mapping description is unfolded in Case Study 1. Furthermore, Fig. 4.8 demonstrates the stability analysis of the distillation column after implementing the 1st and 2nd-order fractional IMC filters. In Fig. 4.8, it is illustrated that all roots lie outside the Riemann sheet, indicating the stability of the distillation column.

Furthermore, the IAE and ISE values presented in Table 4.3 affirm the effectiveness of the recommended controller. The suggested controller consistently demonstrates robust performance even under +10% process parameter deviations.

Figure 4.9 illustrates significant improvements achieved by the proposed controller in both nominal and +10% mismatch scenarios. Figure 4.9a, b highlight that the proposed controller exhibits greater robustness compared to existing controllers, as evidenced by lower absolute sensitivity values.

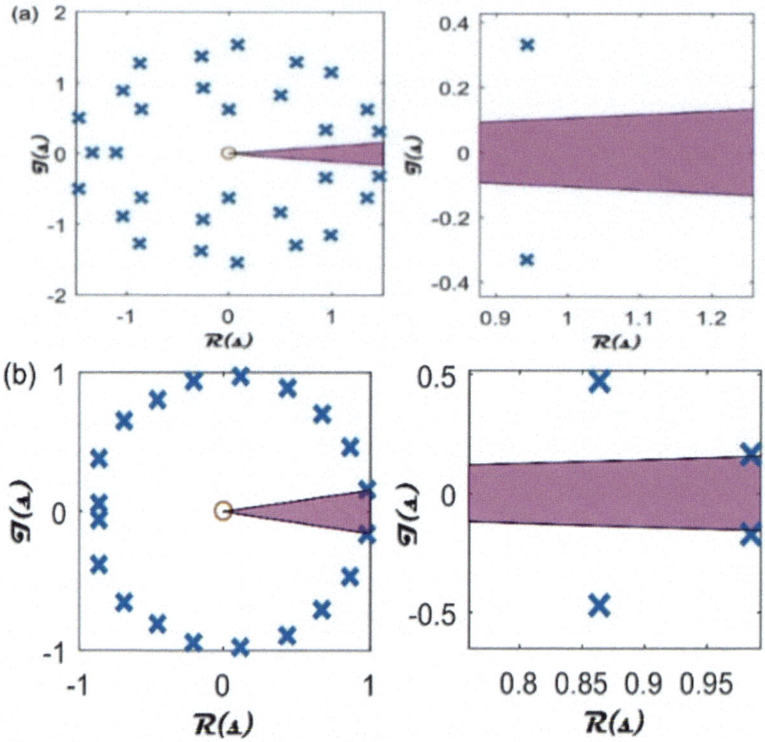

Fig. 4.8 Stability analysis of the distillation column

Table 4.3 Performance indices for the distillation column

Controller	Nominal case		+10% Perturbed case	
	IAE	ISE	IAE	ISE
Proposed controller	0.16	0.026	0.19	0.036
Uma et al. [1]	0.41	0.17	0.50	0.25
Jeng and Lin [5]	0.51	0.26	0.59	0.35
Shamsuzzoha and Lee [6]	1.1	1.3	1.2	1.4

Remark 4.2 In assessing the effectiveness of the suggested controller, comparisons are made with controllers proposed by [1, 5, 6]. These existing controllers lack an inverse response compensator, resulting in higher undershoots, overshoots, and longer settling times compared to the proposed method.

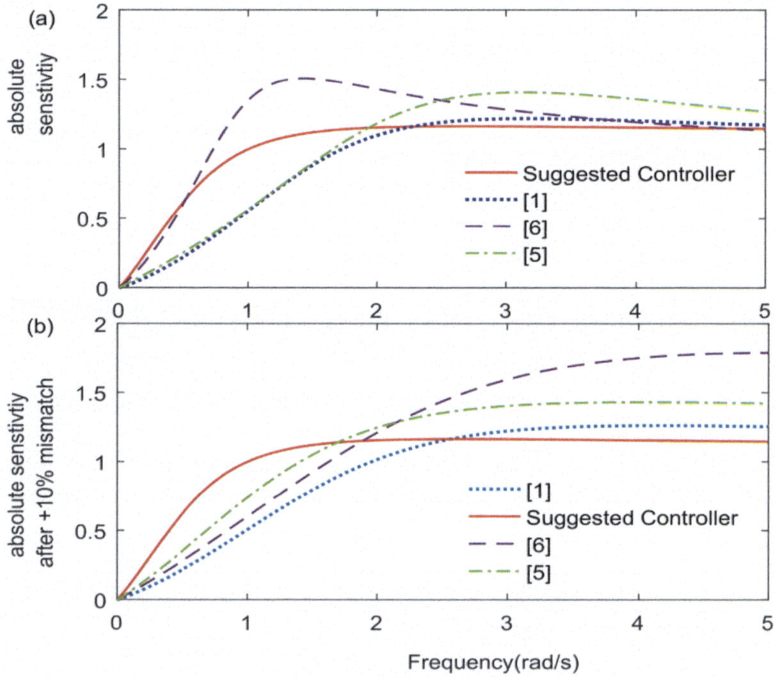

Fig. 4.9 Robustness analysis for the distillation column

4.4.3 Case Study III

The regulation of drum level involves controlling the influence of boiler feed water. According to [2], the transfer function describing the boiler drum level system can be formulated as follows:

$$P_1(s) = \frac{(1 - 0.4699s)}{s} e^{-0.50s},\ P_2(s) = \frac{0.5183}{1.1609s + 1} e^{-0.20s}.$$

Inner loop controller design: For the inner loop, controller settings utilize Eq. (4.5). It can be written as,

$$C_2(s) = \frac{1.1609s + 1}{0.5183(0.95s + 1)}.$$

Outer loop controller design: After the utilization of Eq. (4.6), the overall plant transfer function can be obtained as,

$$\tilde{P}(s) = \frac{(1 - .4699s)(1 + 1.16s)e^{-0.5s}}{0.5183(0.95s + 1)s}. \tag{4.14}$$

The suggested controller settings for the outer loop in the fractional IMC framework can be obtained via Eq. (4.7),

$$C_{f_2}(s) = \frac{0.95s^2 + s}{0.5183(0.004s^3 + 0.08s^2 + 2.1s^{1.3} - 0.624s + 2.1s^{0.3} - 0.720)}\left(0.15 + \frac{1}{s} + 0.008s\right). \tag{4.15}$$

where the term $\frac{0.95s^2+s}{0.5183(0.004s^3 + 0.08s^2 + 2.1s^{1.3} - 0.624s + 2.1s^{0.3} - 0.720)}$ shows fractional filter, and the term $\left(0.15 + \frac{1}{s} + 0.008s\right)$ exhibits PID controller with $PM = 69$, $\lambda = 0.3$, $\tau = 0.2$ and $\phi = 2.1$.

The suggested other controller settings for the outer loop in the fractional IMC framework with a second-order fractional IMC filter can be obtained via Eq. (4.8) (Table 4.4).

$$C_{f_2}(s) = \frac{0.190s^3 + 1.15s^2 + s}{1.8(0.0024s^4 + 0.0084s^3 - 0.42s^2 - 0.718s - 1.17 + 4.4s^{2.60} + 4.2s^{1.2} + 4.4s^{14} + 4.2s^{0.2})}\left(0.15 + \frac{1}{s} + 0.008s\right). \tag{4.16}$$

Table 4.2 indicates that the proposed controller achieves lower values of IAE and ISE.

Figure 4.10 illustrates the square response of the cascade system with a disturbance magnitude of 0.2 at t = 50 s, depicted for both nominal and +10% mismatch scenarios.

Additionally, the stability analysis of the proposed scheme relies on the Riemann surface, which is graphically evaluated using the FOMCON toolbox in MATLAB. The general expression for stability analysis of the first-order fractional IMC filter can be formulated as follows [2]:

Table 4.4 Performance indices for the boiler drum level system

Controller	Nominal case		+10% Perturbed case	
	IAE	ISE	IAE	ISE
Proposed controller	0.16	0.028	0.23	0.056
Uma et al. [1]	0.2	0.04	0.3	0 0.09
Nagarsheth and Sharma [2]	0.3	0010 0.10	0.38	0.14
Raja and Ali [3]	0.4	0.16	0.43	0.18

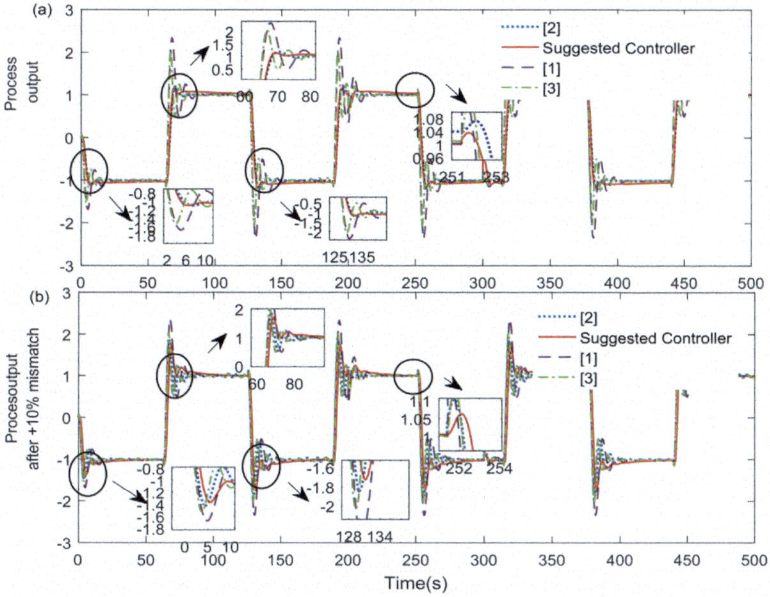

Fig. 4.10 Comparative analysis for boiler drum level system

$$f_2(s) = 1 + \frac{t_{d1}^2 a s^4}{12} + s^3\left(\left(\frac{t_{d_1}a}{2} - \frac{t_{d1}^2}{12}\right) + \frac{t_{d1}^2 a}{12}\right)$$
$$+ s^2\left(\left(\frac{t_{d_1}a}{2} - \frac{t_d^2}{12}\right) - \left(\frac{t_{d_1}}{2} + a\right)\right) + \phi s^{\lambda+2} + \phi s^{\lambda+1} - s\left(\frac{t_{d_1}}{2} + a\right).$$

Likewise, the generalized expression for stability analysis with 2nd order fractional IMC the filter is given as,

$$f_2'(s) = 1 + \frac{t_{d1}^2 a\tau s^5}{12} + s^4\left(\left(\frac{t_{d_1}\tau a}{2} - \frac{t_{d1}^2}{12} + \frac{t_d^2 a}{12}\right) + \frac{t_{d1}^2 a\tau}{12}\right)$$
$$+ s^3\left(\left(\frac{t_{d_1}a}{2} - \tau a - \frac{t_{d_1}\tau}{2} - \frac{t_{d1}^2}{12}\right) + \left(\frac{t_{d_1}\tau a}{2} + \frac{t_d^2\tau}{12} - \frac{t_d^2}{12}\right)\right)$$
$$+ s^2\left(\left(\frac{t_{d_1}a}{2} - \tau a - \frac{t_{d_1}\tau}{2} - \frac{t_{d1}^2}{12}\right) - \left(\tau + a + \frac{t_{d_1}}{2}\right)\right)$$
$$+ \phi^2 s^{2\lambda+3} + 2\phi s^{\lambda+2} + \phi^2 s^{2\lambda+2} + 2\phi s^{\lambda+1} s\left(\frac{t_{d1}}{2} + a + \tau\right).$$

The specific structure of fractional characteristic polynomial for boiler drum level system can be written as,

$$f_2(s) = 1 + 0.002s^4 + 0.043s^3 - 0.312s^2 - 0.360s + 0.5s^{2.3} + 0.5s^{1.3}.$$

$$f_2'(s) = 1 + 0.0014s^5 + 0.004s^4 - 0.021s^3 - 0.359s^2 - 0.585s$$
$$+2.25s^{3.60} + 2.1s^{2.40} + 4.4s^{2.2} + 4.2s^{1.2}.$$

The roots of fractional-order characteristic polynomials lie in the $s = \alpha + j\omega$. The mapping of $s - plane$ a plane linked with an equivalent characteristic polynomial. Choose $\alpha = s^{\beta}$, then linked characteristic polynomial becomes,

$$f_2(\alpha^{1/\beta}) = 1 + 0.002\alpha^{4000} + 0.043\alpha^{3000} - 0.312\alpha^{2000}$$
$$- 0.3603\alpha^{1000} + 0.50\alpha^{2300} + 0.5\alpha^{1300}.$$

$$f_2'(\alpha^{1/\beta}) = 1 + 0.00014\alpha^{5000} + 0.0044\alpha^{4000} - 0.021\alpha^{3000} - 0.359\alpha^{2000}$$
$$- 0.585\alpha^{1000} + 2.25\alpha^{3600} + 2.1\alpha^{2400} + 4.4\alpha^{2200} + 14.2\alpha^{1200}.$$

Where $s^{0.001} = \alpha$. From the stability point of view of the recommended fractional closed-loop system, the absolute values of the angles of roots of associated characteristics polynomial should be more significant $\frac{\beta\pi}{2}$.

The graphical analysis is depicted in Fig. 4.11. Since all the roots lie outside the Riemann sheet, it confirms the stability of the system [2].

The sensitivity analysis confirms the attenuation of disturbances occurring at the sensitivity function $|s(j\omega)| \leq 1$ and the amplification of disturbances occurring $|s(j\omega)| \geq 1$. The sensitivity function is given by [3],

$$S(s) = \frac{1}{1 + \tilde{P}(s)C(s)}. \tag{4.17}$$

Figure 4.12 illustrates a graphical representation of the absolute sensitivity. The maximum absolute sensitivity serves as a reliable measure of robustness, where a lower value indicates greater robustness.

Remark 4.3 To assess the effectiveness of the proposed controller, it is compared with the approaches in [1–3]. Their methods employ lower-order filters and approximations for controller design. Specifically, [1] utilizes a lower-order filter along with a setpoint filter. In contrast, this study employs a higher-order fractional IMC filter, eliminating the need for a setpoint filter. In [2, 3], inverse response and dead-time compensators were not included in the controller design.

4.4.4 Case Study IV

The transfer function describing the distillation column is detailed in [3].

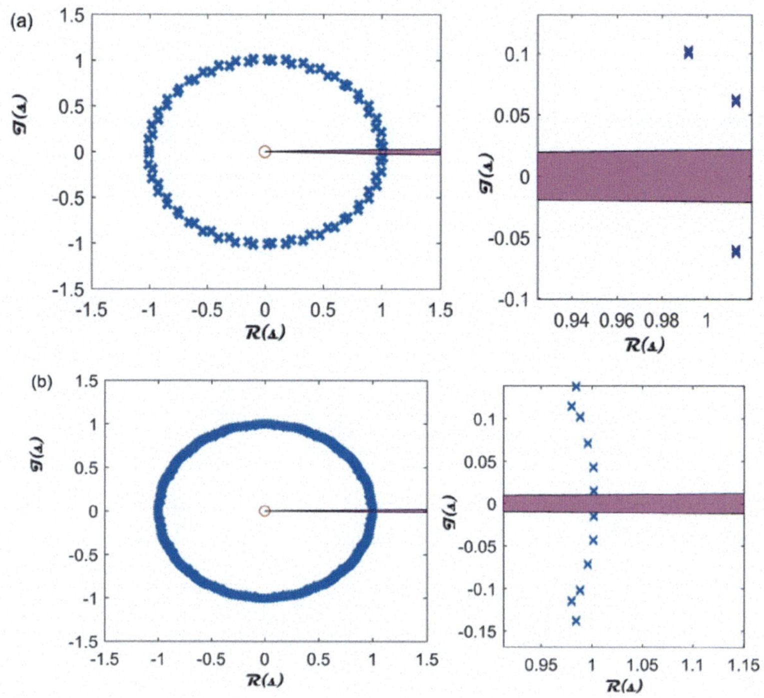

Fig. 4.11 Stability analysis of the boiler drum level system

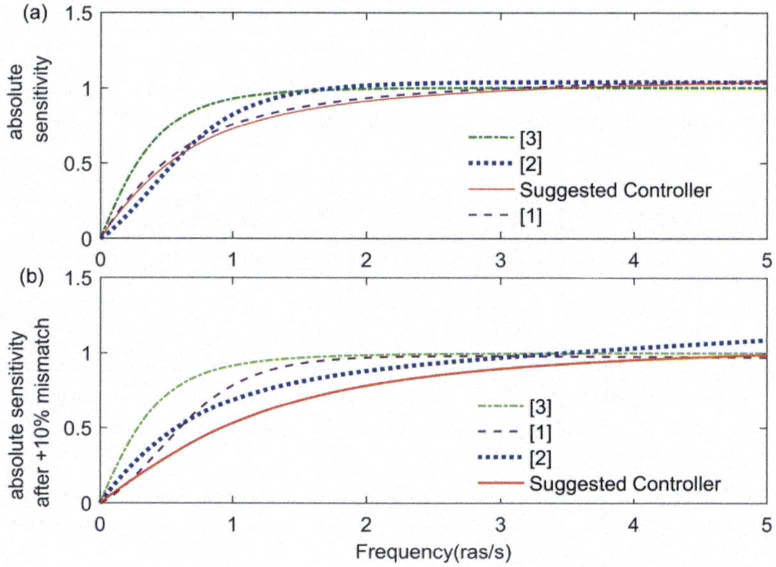

Fig. 4.12 Robustness analysis for the boiler drum level system

$$\tilde{P}_1(s) = \frac{(1 - 1.62s)}{2.71s + 1}e^{-0.4s}, \tilde{P}_2(s) = \frac{0.94}{1.50s + 1}e^{-0.2s}.$$

Inner loop controller design: The inner controller is obtained from Eq. (4.5). It can be written as,

$$C_2(s) = \frac{1.50s + 1}{0.94(0.6s + 1)}.$$

After the utilization of Eq. (4.11), the overall plant transfer function is modified as,

$$\tilde{P}(s) = \frac{(1 - 0.62s)(1 + 1.50s)e^{-0.2s}}{0.94(0.6s + 1)(2.71s + 1)}. \tag{4.18}$$

Outer loop controller design: The suggested controller settings for the outer loop in the fractional IMC framework can be obtained via Eq. (4.12). It can be written as,

$$C_{f_2}(s) = \frac{1.62s^2 + 3.3s + 1}{(0.008s^3 + 0.24s^2 - 2.57s - 1.72 + s^{1.10} + ss^{0.10})}\left(0.05 + \frac{1}{s} + 0.0033s\right). \tag{4.19}$$

where the term $\frac{0.3s^2+1.3s+1}{(0.0033s^3+0.099s^2-0.29s-0.3+0.65s^{1.27}+0.65s^{0.27})}$ exhibits a fractional filter $\lambda = 0.1$, $\tau = 0.1$ and $\phi = 1$. The term $\left(0.05 + \frac{1}{s} + 0.0033s\right)$ displayed display PID controller.

The suggested other controller settings for the outer loop in fractional IMC framework with second-order fractional IMC filter can be obtained via Eq. (4.13).

$$C_{f_2}(s) = \frac{0.12s^3 + 0.82s^2 + 1.7s + 1}{(0.00007s^4 + .00337s^3 - 0.04s^2 - 0.70s - 0.55 + 0.42s^{2.54} + 1.30s^{1.27} + 0.42s^{1.54} + 1.30s^{0.27})}\left(0.05 + \frac{1}{s} + 0.0033s\right) \tag{4.20}$$

In Fig. 4.13a, b, it is evident that the proposed controller exhibits quicker settling times following disturbances with sawtooth input. Figure 4.13 illustrates that the settling performance of the proposed controller yields significant improvements in outcomes.

The description of the fractional characteristic polynomial for the distillation column after utilizing 1st order fractional IMC filter and 2nd-order Pade approximation can be written as,

$$f_2(s) = 1 + 0.008s^4 + 0.24s^3 - 2.5s^2 - 1.72s + s^{2.10} + s^{1.10}.$$

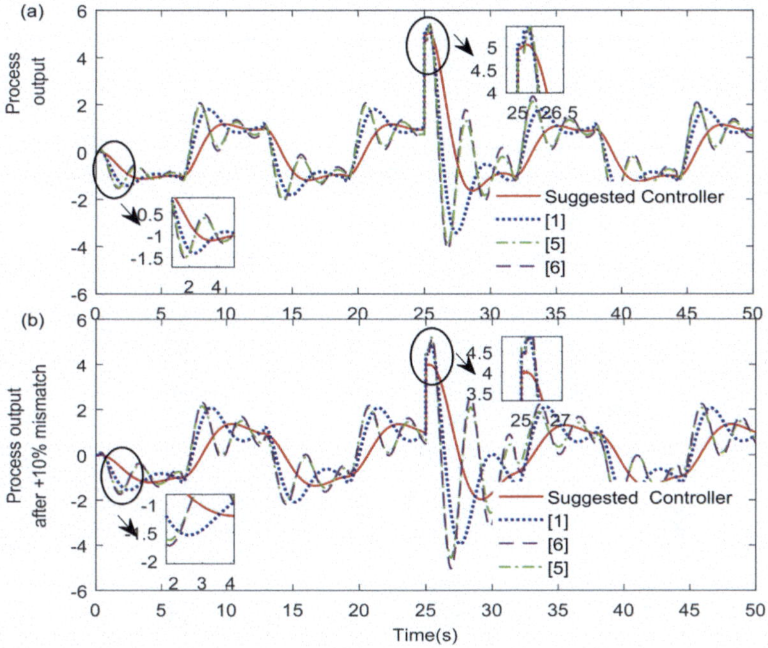

Fig. 4.13 Relative investigation study for the distillation column

Likewise, fractional characteristic polynomial for the distillation column after utilizing 2nd order fractional IMC filter and 2nd order Pade approximation can be written as,

$$f_2(s) = 1 + 0.010s^5 + 0.02s^4 - 0.00027s^3 - 2.74s^2 - 1.82s + 1.5s^{3.20} + 3s^{2.10} + 2.42s^{2.20} + 2s^{1.10}.$$

The detailed mapping description is unfolded in Case Study 1. Furthermore, Fig. 4.14 demonstrates the stability analysis of the distillation column after implementing the 1st and 2nd-order fractional IMC filters. In Fig. 4.14, it is illustrated that all roots lie outside the Riemann sheet, indicating the stability of the distillation column.

Furthermore, the IAE and ISE values presented in Table 4.5 affirm the effectiveness of the recommended controller. The suggested controller consistently demonstrates robust performance even under +10% process parameter deviations.

Figure 4.15 illustrates significant improvements achieved by the proposed controller in both nominal and + 10% mismatch scenarios. Figure 4.15a, b highlight that the proposed controller exhibits greater robustness compared to existing controllers, as evidenced by lower absolute sensitivity values.

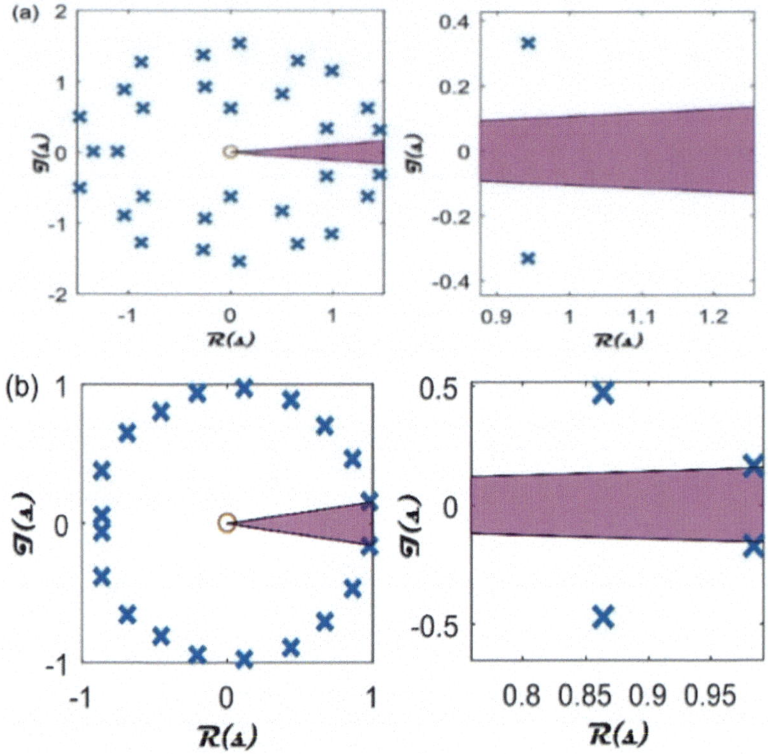

Fig. 4.14 Stability examination of the distillation column

Table 4.5 Performance analysis of the distillation column

Controller	Nominal case		+10% Perturbed case	
	IAE	ISE	IAE	ISE
Proposed controller	0.16	0.026	0.19	0.036
Uma et al. [1]	0.41	0.17	0.50	0.25
Shamsuzzoha and Lee [6]	0.51	0.26	0.59	0.35
Jeng and Lin [5]	1.1	1.3	1.2	1.4

Remark 4.4 In assessing the effectiveness of the suggested controller, comparisons are made with controllers proposed by [1, 6, 5]. These existing controllers lack an inverse response compensator, resulting in higher undershoots, overshoots, and longer settling times compared to the proposed method.

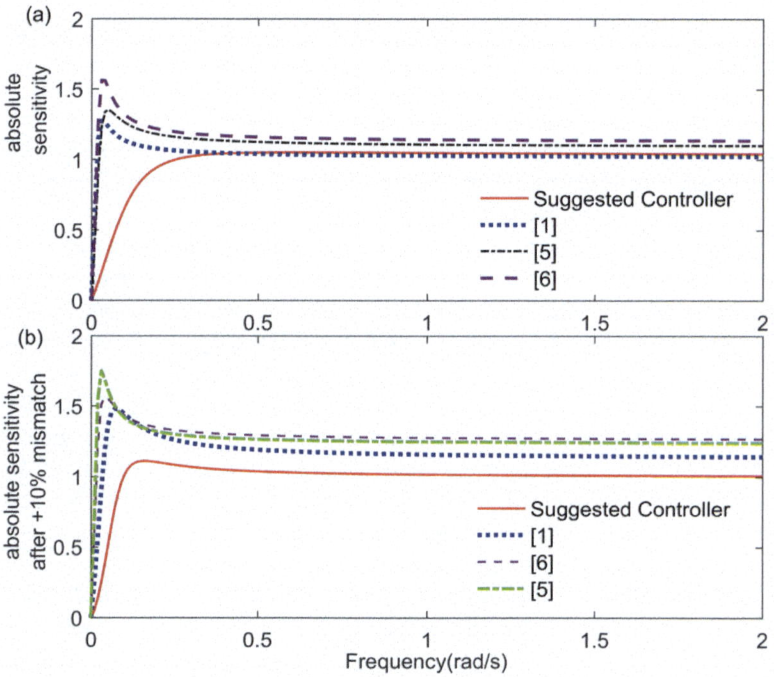

Fig. 4.15 Robustness investigation for the distillation column

4.5 Summary

This chapter introduces the parallel cascade control scheme, which is presented following a discussion on the accountability of inverse response and the dead-time compensator. Despite its advantages, the parallel cascade control scheme faces challenges in achieving an optimal closed-loop response when dealing with Non-Minimum Phase (NMP) zeros and dead time. These issues are explored in more depth in Chap. 5, where the limitations and potential improvements of the scheme are discussed.

4.6 Some Informative Views of Chapter in Post-mortem Form

Question-1: What is parallel cascade control, and how does it differ from traditional single-loop control systems.

Answer: *Parallel cascade control is a sophisticated control strategy used in process control systems to enhance stability and performance. It involves multiple control*

loops working together to regulate a process more effectively. Here's a breakdown of how it works:

Basic Concept

- *Primary Control Loop (Master Loop): This is the main control loop that regulates the primary variable of interest. For example, in a temperature control system, this loop might control the temperature of a furnace.*
- *Secondary Control Loops (Slave Loops): These loops work in parallel with the primary loop and are used to manage secondary variables that influence the primary variable. Continuing with the temperature control example, a secondary loop might control the flow rate of a heating fluid that affects the temperature.*

How It Works

- *Cascade Control: In a typical cascade control scheme, there's a hierarchical relationship where one controller (the primary or master controller) sets the setpoint for another controller (the secondary or slave controller). The secondary controller then acts to minimize the deviation of its controlled variable, which in turn helps maintain the primary variable closer to its setpoint.*
- *Parallel Operation: In a parallel cascade control scheme, multiple secondary control loops operate in parallel rather than in a strictly hierarchical manner. This means each secondary loop works independently to regulate its respective variable, but its outputs are used to influence the primary loop.*

Advantages

- *Improved Performance: By using multiple control loops in parallel, the system can handle disturbances and variations more effectively. This leads to better overall performance and stability.*
- *Reduced Interactions: It can help reduce the interaction between different control loops and minimize the effect of disturbances on the primary variable.*
- *Enhanced Precision: With parallel control loops, each loop can focus on a specific aspect of the process, leading to finer control and improved accuracy.*

Applications

Parallel cascade control is used in various industries, including chemical processing, manufacturing, and energy systems, where precise control of multiple variables is essential for optimal operation.

In essence, the parallel cascade control scheme leverages the strengths of multiple control loops operating simultaneously to achieve better control over complex processes.

Question-2: How does parallel cascade control contribute to improving the performance, stability, and robustness of control systems?

Answer: *The parallel cascade control offers a more sophisticated and robust control strategy compared to single-loop control systems, enabling better performance, stability, and adaptability in industrial processes.*

Question-3: Can you describe a scenario where parallel cascade control would be particularly advantageous over other control strategies?

Answer: *The parallel cascade control would be particularly advantageous in temperature control applications for chemical reactors due to its ability to handle complex process dynamics, provide fast disturbance rejection, improve stability and robustness, ensure tight control over product quality, and adapt to changes in process conditions effectively.*

Question-4: In what types of industrial processes is parallel cascade control commonly used, and why?

Answer: *The parallel cascade control is widely used in industries characterized by complex processes with multiple interacting variables. Its ability to simultaneously regulate primary and secondary variables enables precise control, improved process performance, and enhanced product quality, making it a preferred control strategy in various industrial applications.*

Question-5: How inverse response and dead-time compensator is useful with parallel cascade control

Answer: *The inverse response and dead-time compensators play crucial roles in parallel cascade control by addressing challenging process dynamics and improving control system performance. By effectively compensating for inverse response behavior and dead time, these compensators contribute to the stability, robustness, and responsiveness of parallel cascade control systems, enabling more accurate and reliable control of industrial processes.*

Question-6: What are the implementation issues in cascade control?

Answer: *Cascade control is a technique used in control systems to improve the performance and stability of a process by using multiple feedback loops. While it can be very effective, its implementation can come with several issues. Here are some common challenges:*

- ***Tuning Complexity***: *Cascade control involves tuning multiple controllers—both the primary (outer) controller and the secondary (inner) controller. Each controller affects the other, making the tuning process more complex than with a single-loop system. Properly setting the proportional, integral, and derivative (PID) parameters for each controller is crucial and often requires iterative adjustments and expertise.*
- ***Interaction Between Loops***: *In cascade control, the inner loop is typically faster and has a shorter time constant than the outer loop. If the interactions between*

the inner and outer loops are not managed correctly, it can lead to oscillations or instability. Ensuring that the inner loop responds quickly enough without causing adverse effects on the outer loop is important.

- ***Dynamic Modeling***: *Accurate modeling of the process dynamics is essential for effective cascade control. If the model used to design the controllers is not representative of the actual process dynamics, the controllers may not perform as intended. Changes in the process or disturbances can further complicate this issue.*
- ***Sensor and Actuator Dynamics***: *The performance of cascade control systems depends on the quality of sensors and actuators. If these components are not responsive or accurate, the effectiveness of the cascade control can be compromised. Sensor noise or actuator lag can lead to poor control performance.*
- ***Cascade Loop Design***: *Deciding which variables to control with the inner and outer loops can be challenging. The inner loop typically controls a variable that directly affects the output of the process, while the outer loop controls the overall process variable. Incorrectly choosing or designing these loops can lead to suboptimal performance.*
- ***Stability Concerns***: *Ensuring stability in cascade control systems can be difficult, especially if the inner and outer loops are not properly tuned. Instability can result from improper loop interactions or the inner loop having too much influence on the outer loop.*
- ***Implementation Costs***: *Implementing a cascade control system often requires additional hardware and software compared to a single-loop system. This can increase both the initial cost and the complexity of the system, including the need for more sophisticated controllers and tuning tools.*
- ***Maintenance and Calibration***: *Cascade control systems may require more frequent maintenance and calibration compared to simpler control systems. The need to regularly tune both loops to adapt to changing process conditions or disturbances can increase the maintenance burden.*
- ***Complexity of Troubleshooting***: *Diagnosing issues in a cascade control system can be more complex due to the interactions between multiple controllers. Problems in the inner loop can affect the outer loop and* vice versa, *making it harder to pinpoint the root cause of issues.*

References

1. Uma S, Chidambaram M, Rao AS, Yoo CK (2010) Enhanced control of integrating cascade processes with time delays using modified Smith predictor. Chem Eng Sci 65(3):1065–1075
2. Nagarsheth SH, Sharma SN (2020) Control of non-minimum phase system with dead time: a fractional system viewpoint. Int J Syst Sci 51(11):1905–1928
3. Raja GL, Ali A (2021) Enhanced tuning of Smith predictor-based series cascade control structure for integrating. ISA Trans 114:191–205
4. Pashaei S, Bagher P (2020) Parallel cascade control of dead time process via fractional order controller based on smith predictor. ISA Trans 98:186–197

5. Jeng JC, Lin SW (2012) Robust proportional-integral derivative controller design for stable/integrating processes with an inverse response and time delay. Ind Eng Chem Res 51(6):2652–2665
6. Shamsuzzoha M, Lee M (2008) Design of advanced PID controller for enhanced disturbance rejection of second-order processes with time delay. AIChE J 54(6):1526–1536
7. Joshi D, Deb D, Giri AK (2024) MRAS disturbance observer-based sensorless field-oriented backstepping control of BLDC motor drive. In: Electrical engineering. Springer, Berlin. https://doi.org/10.1007/s00202-024-02378
8. Kundu S, Singh M, Giri AK (2022) Implementation of in-loop filter based DSC-FLL control approach for power quality improvement in isolated microgrid. In: Proceedings of 2022 IEEE 31st international symposium on industrial electronics (ISIE). USA. https://doi.org/10.1109/ISIE51582.2022.9831712
9. Giri AK, Arya SR, Maurya R (2020) Hybrid order generalized integrator based control for VSC to improve the PMSG operation in isolated mode. In: The proceedings of 2020 first international conference on power, control and computing technologies (ICPC2T). India. https://doi.org/10.1109/ICPC2T48082.2020.9071498
10. Kundu S, Singh M, Giri AK (2024) Synchronization and control of WECS-SPV-BSS-based distributed generation system using ICCF-PLL control approach. Electr Power Syst Res 226:109919
11. Giri AK, Arya SR, Guerrero JM, Kumar S, Power quality: infrastructures and control. Springer Nature, Singapore. ISBN: 978-981-19-7956-9
12. Giri AK, Arya SR, Vinnikov D, Distributed energy systems design, modeling, and control, vol 1. CRC Press, pp 1–308. ISBN: 9781032134246
13. Arya SR, Patel MM, Alam SJ, Srikakolapu J, Giri AK, Chitti Babu B (2021) Classical control algorithms for permanent magnet synchronous generator driven by diesel engine for power quality. Int J Circuit Theory Appl 49(3):576–601
14. Kundu S, Giri AK, Singh M, Kadiyan S, Adaptive control strategy for isolated renewable energy-based generation system with intermittency. In: Electrical engineering. Springer, Berlin. https://doi.org/10.1007/s00202-024-02454-0
15. Meena DC, Singh M, Giri AK (2021) Leaky-momentum control algorithm for voltage and frequency control of three-phase SEIG feeding isolated load. J Eng Res ICARI Spec 109:120
16. Meena DC, Singh M, Giri AK (2021) A modified NLMS control algorithm for coordinated operation in three-phase wind-energy conversion system. Int J Renew Energy Res (IJRER) 11(4):1621–1629
17. Dalwadi N, Deb D, Ozana S (2023) Performance evaluation of RW-quadrotor and Bi-quadrotor for payload delivery. IET Intel Transp Syst 17(11):2221–2236
18. Dalwadi N, Deb D, Ozana S, Adaptive hybrid control of quadrotor drones. 461:1–174. ISBN: 978-981-19-9743-3

Chapter 5
Modified Tuning Approaches for Non-minimum Phase Systems

5.1 Introduction

The effectiveness of cascade control in process industries is greatly influenced by how well the system is tuned. This tuning process, however, can become quite intricate when Non-Minimum Phase (NMP) zeros and dead time are present in the process transfer function.

5.2 Modified Tuning Approach for Fractional-Order Filter

Cascade control is an advanced and responsive method compared to a single feedback loop, incorporating flow, level, and pressure control loops to mitigate external disturbances. This approach features both inner and outer loops [1]. The success of cascade control hinges on the precise tuning of both loops. PID controllers are often employed due to their flexibility and the range of available tuning methods [2]. The Internal Model Control (IMC) technique is a widely used method for tuning cascade systems [2, 3]. Enhancements in tuning for Smith predictor-based series cascade control, particularly for integrating processes, have been explored [4], and issues related to delays have been discussed [5].

Existing methods include tuning using fractional-order filters with traditional and delayed Bode's ideal transfer functions. This chapter focuses on tuning the outer loop controller using the delayed Bode's ideal transfer function within the IMC framework.

M. Yadav et al., *Control Strategies for Non-minimum Phase Systems*,
Studies in Infrastructure and Control, https://doi.org/10.1007/978-981-96-2351-8_5

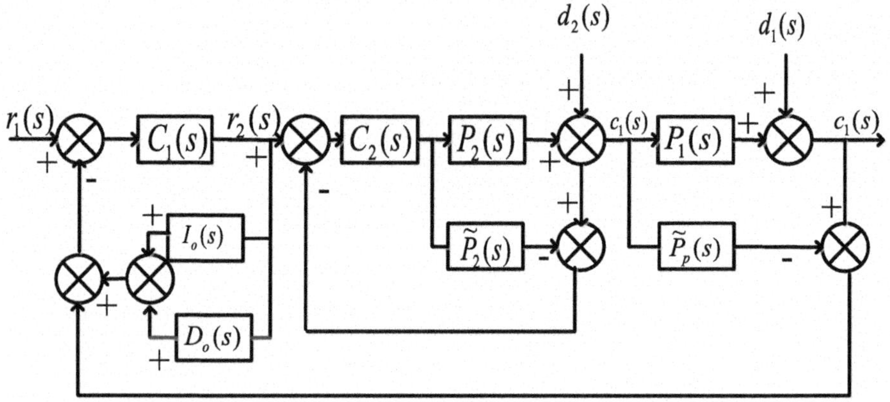

Fig. 5.1 Control arrangement in IMC insight

5.3 Theoretical Development

In Fig. 5.1, the controllers, $C_1(s)$, $C_2(s)$ are termed master and slave controllers, respectively.

The transfer function of the inner and outer loop of series cascade control becomes,

$$\frac{c_1(s)}{r_1(s)} = \frac{P_1(s)P_2(s)C_1(s)C_2(s)}{1 + (P_1(s) - \tilde{P}_1(s))P_2(s)C_1(s) + (P_2(s) - \tilde{P}_2(s)C_2(s)}. \tag{5.1}$$

$$\frac{c_1(s)}{d_1(s)} = \frac{(1 - \tilde{P}_1(s)P_1(s)C_1(s))d_1(s) + (1 - \tilde{P}_2(s)C_2(s))P_1(s)d_2(s)}{1 + (P_1(s) - \tilde{P}_1(s))P_2(s)C_1(s) + (P_2(s) - \tilde{P}_2(s)C_2(s)}. \tag{5.2}$$

5.4 Delayed Bode's Ideal Transfer Function Approach

The approach of using the delayed Bode's ideal transfer function in an open-loop framework is termed [5],

$$B'(s) = \frac{1}{\phi s^{\lambda}} e^{-\gamma s}, \lambda \in R. \tag{5.3}$$

The following formula calculates the fractional filter parameters (ϕ, λ),

$$\lambda = \frac{\pi - P.M. - \omega t_d}{\pi/2}, \phi = (\omega)^{-\lambda}. \tag{5.4}$$

5.5 Suggested Work

The objective of this section is to fine-tune the fractional filter parameters of the outer loop controller by applying a modified Bode's ideal transfer function. Thus, a delayed version of Bode's ideal transfer function is used for this purpose [5].

Inner loop controller design: In the context of the IMC framework, the transfer function for the inner loop controller is expressed as [1],

$$C_2(s) = \frac{f(s)}{\tilde{P}_2^-(s)}, \tag{5.5}$$

Based on the examination of Fig. 5.1, the outcome for the inner loop is

$$H(s) = \frac{C_2(s)\tilde{P}_2(s)}{1 + C_2(s)(P_2(s) - \tilde{P}_2(s))}. \tag{5.6}$$

Outer loop controller design: Initially, we aimed to determine the overall plant transfer function for designing the outer loop controller. This function can be derived from the results of the inner loop and the process transfer function of the outer loop (Figs. 5.2 and 5.3).

$$P(s) = H(s)\tilde{P}_1(s). \tag{5.7}$$

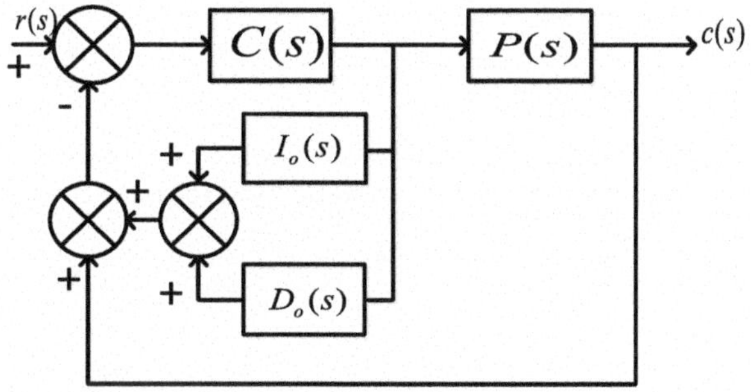

Fig. 5.2 Classical closed-loop control with inverse response compensator

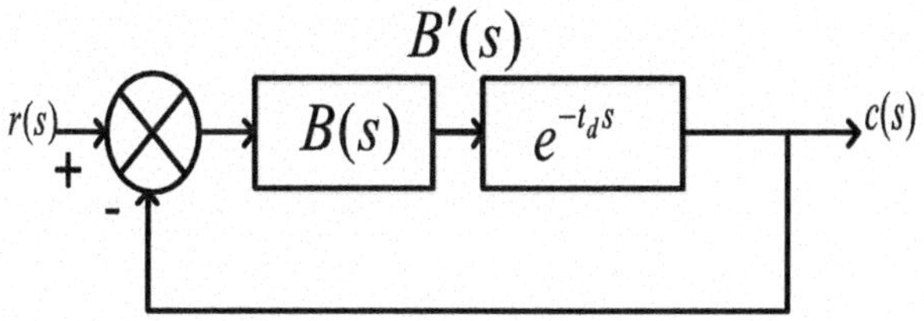

Fig. 5.3 Delayed Bode's ideal transfer function

The inverse response and dead-time compensator are termed as $I_o(s) = 2as\tilde{P}^{-}(s)$ and $D_o(s) = \frac{1-e^{-t_d s}}{e^{-t_d s}}\tilde{P}^{-}(s)$ The controller transfer function from Eq. (5.7) becomes,

$$C(s) = \frac{B'(s)}{P(s)}\left[1 + C(s)\left\{\delta s P(s) + \frac{1-e^{-t_d s}}{e^{-t_d s}}P(s)\right\}\right] \tag{5.8}$$

Equation (5.8) can be modified as,

$$C(s) = \frac{B'(s)}{P(s)\left[1 - B'(s)\left\{\delta s + \frac{1-e^{-t_d s}}{e^{-t_d s}}\right\}\right]} \tag{5.9}$$

Outline of controllers (Precise case 1)

Inner loop controller design: Consider a FOPDT model,

$$\tilde{P}_2(s) = \frac{K_2'}{Ts+1}e^{-t_{d_2}s},$$

where K_2' is the gain, t_{d_2} is the time delay and c is the time constant,
Integer order filter.

$$f_1(s) = \frac{1}{\eta s + 1}.$$

As defined in Eq. (5.5), the transfer function for the inner loop controller is,

$$C_s(s) = \frac{(Ts+1)}{K_2(\eta s + 1)^n}. \tag{5.10}$$

Here n is an integer to make the controller ($C_2(s)$) transfer function proper. Likewise, maximum sensitivity (M_s) can be used for tuning η [1].

Outer loop controller design: Here, integral order plus time delay system is considered as,

$$\tilde{P}_1(s) = \frac{K_1'(1-as)}{s}e^{-t_{d_1}s},$$

where gain, time delay, and NMP zero are denoted by K_1', t_{d_1}, and a. The overall plant transfer function can be obtained via Eq. (5.6),

$$\tilde{P}(s) = \frac{K_1'(1-as)}{s(s+1)}e^{-t_d s} \quad t_d = t_{d_1} + t_{d_2}. \tag{5.11}$$

Following the substitution of the value $\tilde{P}(s)$ from Equation $B'(s)$ (5.11) into Eq. (5.3) and then into Eq. (5.9), we reach Eq. (5.12). This can be formulated as,

$$C(s) = \frac{s}{K_1'\left[\phi s^{\lambda} - (1 - as)(1 - e^{-t_d s})\right]}(1 + \eta s). \tag{5.12}$$

After approximation of the delay term via Taylor series approximation, Eq. (5.12) can be modified as,

$$C(s) = \frac{s^{1-\lambda}}{K_1'\left[\phi s^{\lambda} - (1 - as)(1 - e^{-t_d s})\right]}(1 + \eta s). \tag{5.13}$$

Taylor series approximation $1 - e^{-t_d s} = t_d s$.

Outline of the controller (Precise case 2)

Inner loop controller design: In the aforementioned specific case 1, the secondary loop controller design is outlined.

$$C_2(s) = \frac{(Ts + 1)}{K_2'(\eta s + 1)^n}. \tag{5.14}$$

Outer loop controller design: Here first order plus time delay system is considered as,

$$\tilde{P}_1(s) = \frac{K_1'(1 - as)}{T_1 s + 1} e^{-t_{d_1} s}.$$

From Eq. (5.6), the outer loop process model becomes,

$$\tilde{P}(s) = \frac{K(1 - as)}{(T_1 s + 1)(\eta s + 1)} e^{-t_d s}. \tag{5.15}$$

After utilization Eq. (5.9), the outer loop controller transfer function becomes,

$$C(s) = \frac{s^{1-\lambda}}{K[\phi - t_d s^{1-\lambda} + a t_d s^{2-\lambda}]}\left(\eta T_1 s + (T_1 + \eta) + \frac{1}{s}\right). \tag{5.16}$$

5.6 Particle Swarm Optimization (PSO)

In 1995 Kennedy and Eberhart established the PSO algorithm [6] (Fig. 5.4).

Optimal tuning of the FOPID controller for a single canonical tank system is achieved through the PSO algorithm as proposed [6].

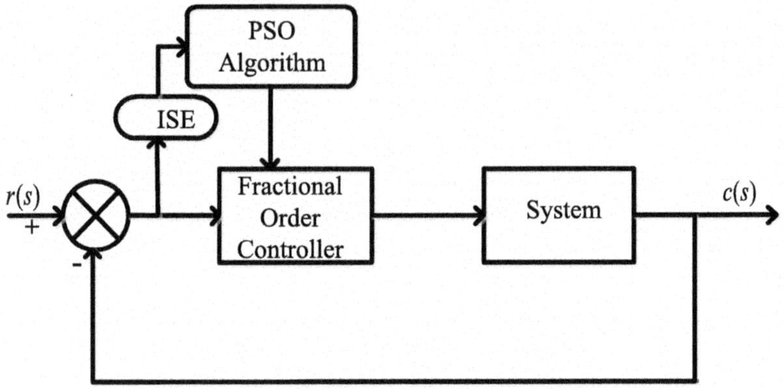

Fig. 5.4 Tuning method of controller parameters via PSO algorithm

5.7 Case Studies

To test the validity of the proposed technique, two case studies are conducted: one with nominal process parameters and another with a +10% mismatch.

5.7.1 Case Study I

The boiler drum level system transfer function can be shown,

$$\tilde{P}_1(s) = \frac{(1 - 0.418s)e^{-0.07s}}{s}, \tilde{P}_2(s) = \frac{0.547e^{-0.03s}}{1.06s + 1}.$$

Inner loop controller design: From Eq. (5.10),

$$C_2(s) = \frac{1.06s + 1}{0.547(0.12s + 1)}.$$

Outer loop controller design: From Eq. (5.11), the outer loop controller can be given as,

$$\tilde{P}(s) = \frac{(1 - 0.418s)e^{-0.1s}}{s(0.12s + 1)}. \tag{5.17}$$

The optimal settings for the outer loop controller in the fractional IMC framework can be derived from Eq. (5.13)

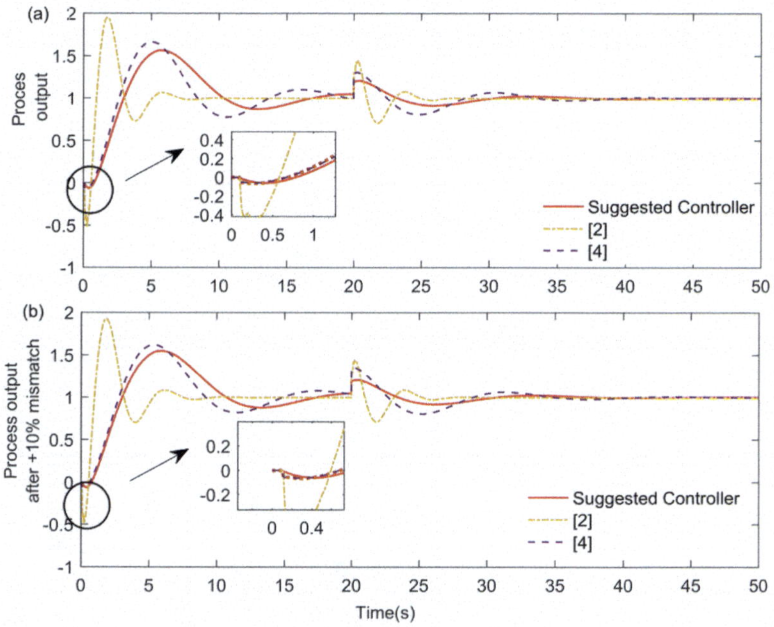

Fig. 5.5 Step response for liquid level system

$$C(s) = \frac{s^{0.10}}{0.65 - 0.936s^{0.10} + 0.0418s^{1.10}}(0.12s + 1), \tag{5.18}$$

where the term $\frac{s^{0.10}}{0.65-0.936s^{0.10}+0.0418s^{1.10}}$ denotes the fractional filter and $(0.12s + 1)$ interprets the PID controller. Here $\eta = 0.12$, $\phi = 0.65$ $\lambda = 0.90$ $PM = 88$. The comparative analysis of the proposed controller is displayed in Fig. 5.5, which also includes the step response of the cascade system subjected to a 0.2 magnitude disturbance at t = 50 s for nominal and +10% mismatch situations.

Table 5.1 demonstrates the IAE and ISE assessments.

For the closed-loop response of the boiler drum level system, the corresponding fractional characteristic polynomial is defined as,

$$f(s) = 1 + 0.65s^{0.10} - 0.1s^{1.10} + 0.0418s^{2.10}.$$

Select $s = j\omega$ for mapping in-plane, i.e., $\alpha = s^{\beta}$ then it can be written as,

$$f(\alpha^{1/\beta}) = 1 + 0.65\alpha^{1000} - 0.1\alpha^{1100} + 0.0418\alpha^{2100}.$$

where $s^{0.001} = \alpha$.

The plot in Fig. 5.6 indicates that all the roots are located outside the Riemann sheet, which demonstrates the stability of the liquid level system [2].

Table 5.1 Performance indices for the liquid level system

Method	Nominal case					+10% mismatch				
	Peak O.S. ratio, %	Peak time, (s)	Rise time, (s)	IAE	ISE	Peak O.S. ratio, %	Peak time, (s)	Rise time (s)	IAE	ISE
Suggested controller	1.23	7.6	4.13	0.19	0.03	1.25	7.5	5.14	0.36	0.28
Nagarsheth and Sharma [2]	1.43	6.25	3.6	0.24	0.12	1.44	4.90	0.5	0.39	0.36
Raja and Ali [4]	1.92	1.94	1.03	0.35	0.37	1.95	1.79	0.98	1.2	0.52

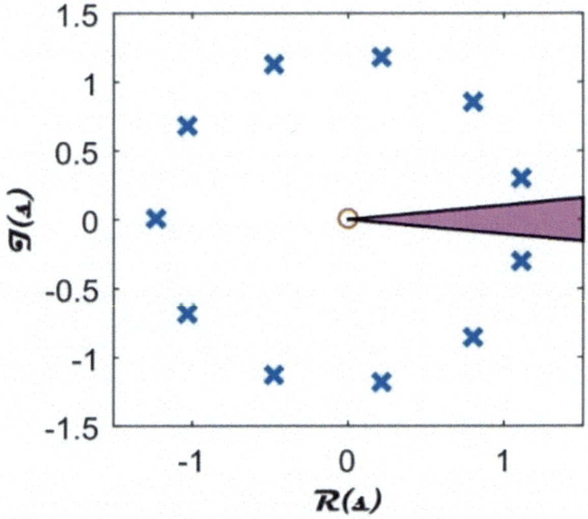

Fig. 5.6 Stability analysis of the liquid level system

Additionally, variations in process parameters influence the stability margin and robustness. The sensitivity function is expressed as [2].

$$S(s) = \frac{1}{1 + \tilde{P}(s)C(s)}. \tag{5.19}$$

To perform the sensitivity analysis, substitute the plant and controller transfer functions into Eq. (5.19) using Eqs. (5.17) and (5.18). Figure 5.7 illustrates the absolute sensitivity for both nominal and +10% mismatch cases.

To optimize the controller settings, we applied Particle Swarm Optimization (PSO). Figure 5.8 illustrates the flow chart and convergence graph associated with the PSO algorithm.

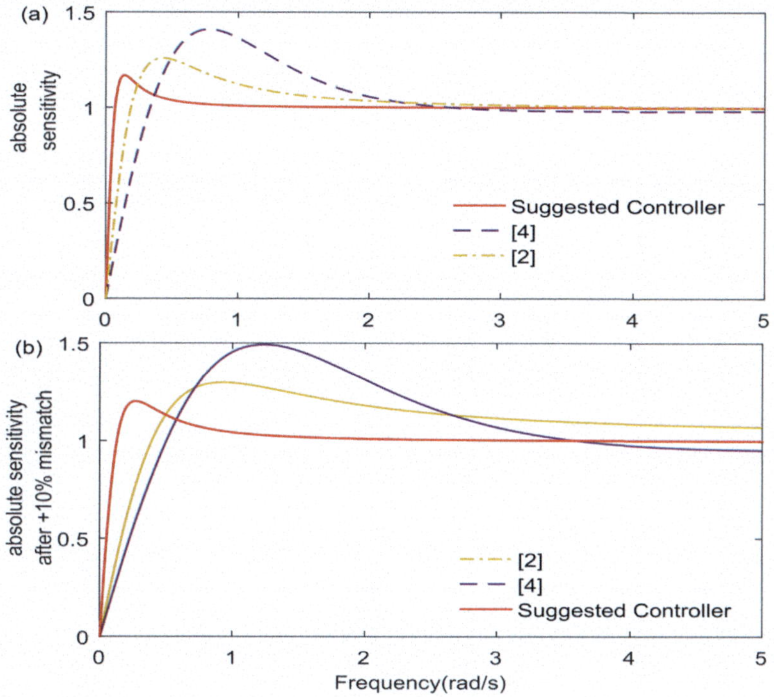

Fig. 5.7 Robustness analysis for the liquid level system

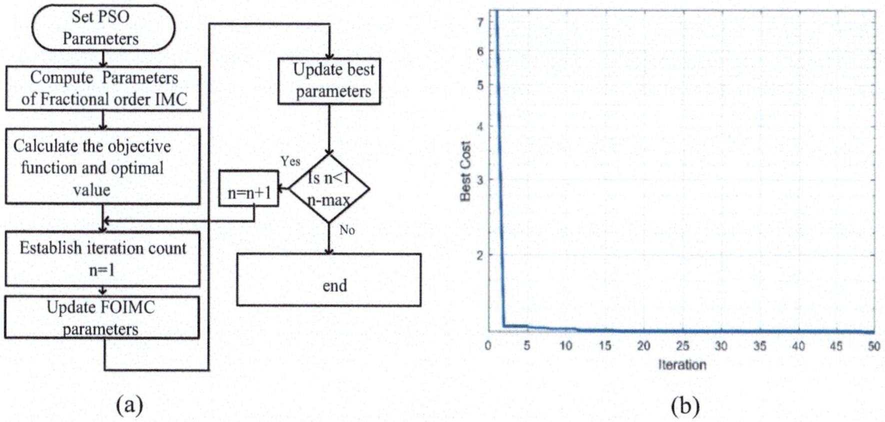

Fig. 5.8 PSO characteristics for boiler drum level system

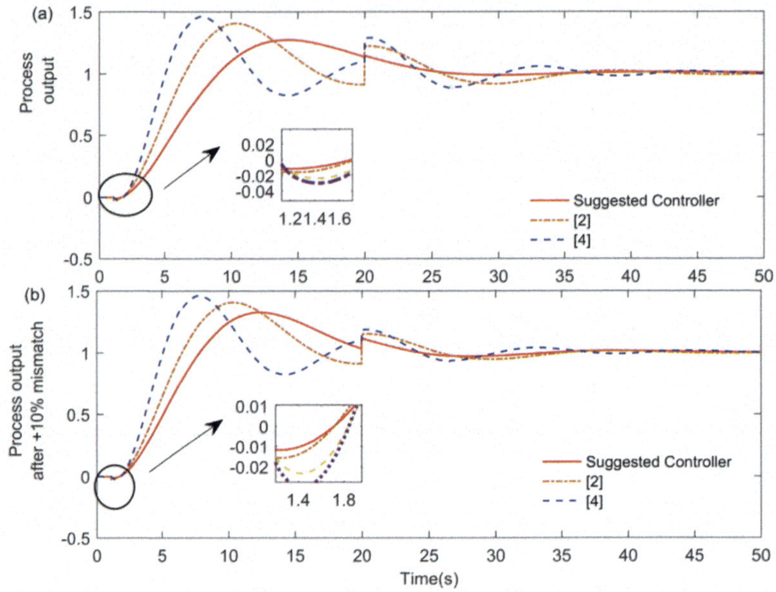

Fig. 5.9 Relative boiler drum level system study using the PSO algorithm

In Fig. 5.9, the outcomes for the boiler drum level system are demonstrated in nominal and +10% mismatch cases after utilization of the PSO algorithm.

Remark 5.1 To gauge the efficacy of the offered controller, it is related to the single feedback loop structures detailed in [2, 4]. The multi-loop controller structure implemented in this work shows a substantial improvement in overshoot, undershoot, and settling time.

5.7.2 *Case Study II*

The transfer function for a higher-order process is specified as given [10],

$$\tilde{P}(s) = \frac{(1 - 2s)e^{-0.6s}}{(2s + 1)(s + 1)(0.5s + 1)}.$$

After being decomposed, the process model is defined as,

$$\tilde{P}_1(s) = \frac{(1 - 1.62s)}{2.71s + 1} e^{-0.4s}, \tilde{P}_2(s) = \frac{0.94}{1.50s + 1} e^{-0.2s}.$$

Inner loop controller design: From Eq. (5.16),

$$C_2(s) = \frac{1.50s + 1}{0.94(0.6s + 1)}.$$

Outer loop controller design: With the implementation of Eq. (5.15), the overall plant transfer function is adapted to,

$$\tilde{P}(s) = \frac{(1 - 1.162s)e^{-0.6s}}{(0.6s + 1)(2.71s + 1)}. \tag{5.20}$$

From Eq. (5.16),

$$C(s) = \frac{s^{0.58}}{0.5 - 0.65s^{0.58} + 1.62s^{1.58}}\left(1.62s + 3.3 + \frac{1}{s}\right), \tag{5.21}$$

where the term $\frac{s^{0.58}}{0.5-0.65s^{0.58}+1.62s^{1.58}}$ specifies fractional IMC filter and $\left(3.33 + 1.62s + \frac{1}{s}\right)$ demonstrates PID controller, the outcomes after using the step signal are revealed in Fig. 5.10a. The impact of a +10% process parameter deviation is shown in Fig. 5.10b.

The stability of higher-order processes can be characterized as

$$f(s) = 1 + 0.5s^{0.42} - 0.6s^{1.42} + 1.62s^{2.42}.$$

Figure 5.11 displays the stability assessment of the higher-order process.

Substituting the value of $\tilde{P}(s)$ and $C(s)$ from Eqs. (5.20) and (5.21) into (5.19).

As illustrated in Fig. 5.12, the suggested controllers show a reduced peak in maximum absolute sensitivity.

The convergence diagram is illustrated in Fig. 5.13. The improved closed-loop response of the controller, achieved with the PSO algorithm, is demonstrated in Fig. 5.14 (Table 5.2).

Remarks 5.2 To assess the effectiveness of the controller, comparisons are made with the approaches outlined in [8, 9]. The existing controllers do not include inverse response and dead-time compensators, whereas this work incorporates a compensator within the outer loop controller and utilizes the delayed Bode's ideal transfer method for fractional filter tuning.

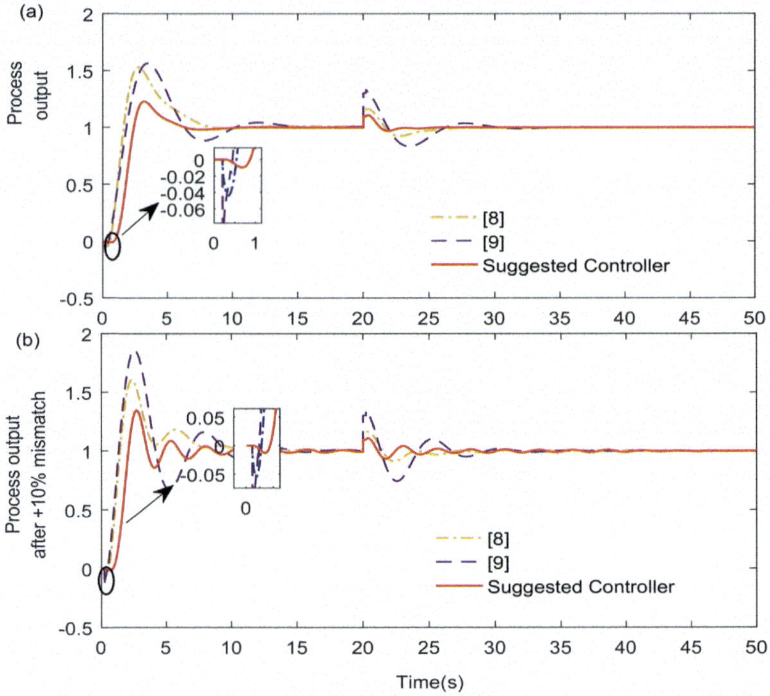

Fig. 5.10 Relative study of the higher-order process

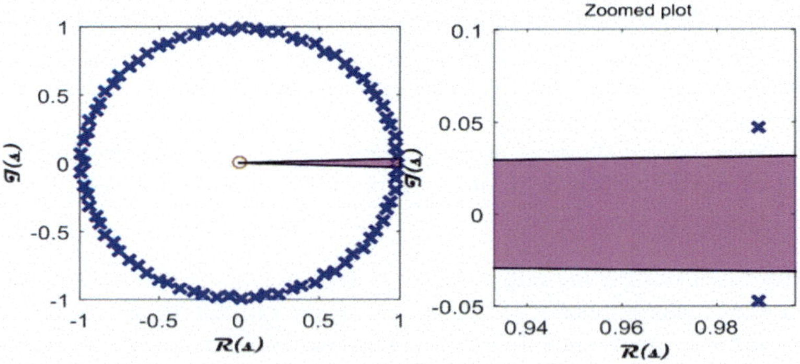

Fig. 5.11 Stability analysis of the higher-order process

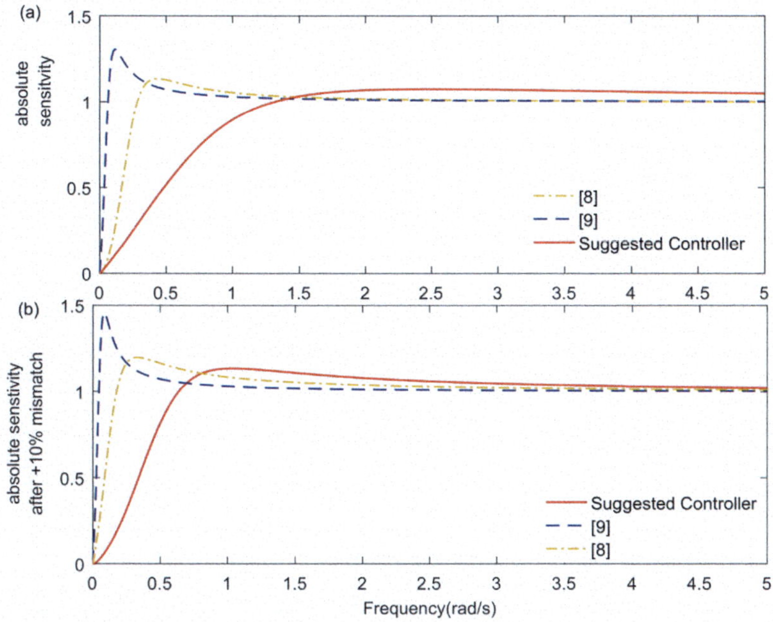

Fig. 5.12 Robustness analysis of the recommended controller

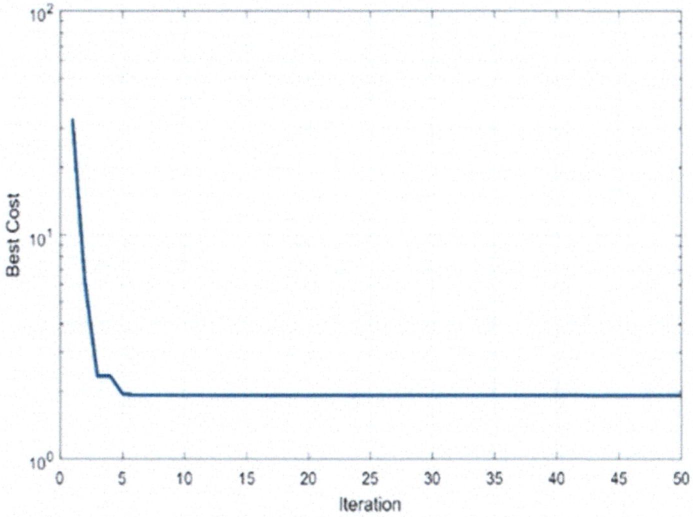

Fig. 5.13 Convergence graph for the higher-order process

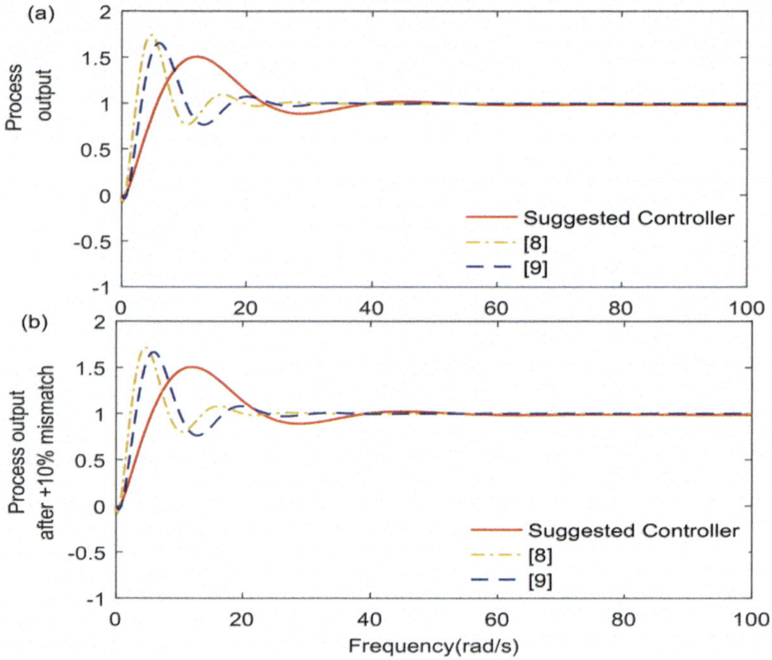

Fig. 5.14 Relative study for higher-order process via PSO algorithm

Table 5.2 Transient analysis and performance measure for the higher-order process

Method	Nominal case					+10% mismatch				
	Peak O.S. ratio, %	Peak time, (s)	Rise time, (s)	IAE	ISE	Peak O.S ratio, %	Peak time, (s)	Rise time (s)	IAE	ISE
Suggested controller	1.12	13.63	10.1	3.7	14.39	1.13	13.37	10.2	3.8	14.58
Pai et al. [8]	1.39	7.5	5.7	5.9	35.95	1.47	10.37	7.6	5.9	23.48
Jeng and Lin [9]	1.22	9.9	7.52	5.6	29.53	1.26	8.3	5.9	4.85	26.29

5.8 Summary

In this chapter delayed Bode's ideal transfer function is used for NMP zeros and dead-time system. However, after encountering the NMP zeros this tuning approach shows some constraints. Thus, in Chap. 6. Bode 's ideal transfer after accountability of NMP zeros.

5.9 Some Informative Views of Chapter in Post-mortem Form

Question 1: What is the importance of tuning of controller?

Answer:

1. *Achieving Desired Performance*
   - *Response Time: Tuning helps achieve the desired speed of response. Properly tuned controllers ensure that the system responds quickly to changes or disturbances.*
   - *Stability: It ensures that the system remains stable under various operating conditions. Poorly tuned controllers can lead to oscillations or even instability.*
   - *Accuracy: Tuning helps minimize the steady-state error, ensuring that the system output closely matches the desired setpoint.*
2. *Improving System Efficiency*
   - *Energy Consumption: Well-tuned controllers can optimize energy usage, reducing unnecessary power consumption or wear and tear on system components.*
   - *Resource Utilization: Proper tuning can lead to more efficient use of system resources, including computational and mechanical resources.*
3. *Enhancing System Robustness*
   - *Disturbance Rejection: A well-tuned controller can better handle disturbances and variations in system dynamics, maintaining performance despite external changes.*
   - *Noise Reduction: Proper tuning can help the system deal with noise effectively, ensuring that performance is not adversely affected by random fluctuations.*
4. *Reducing Oscillations and Over-Response*
   - *Damping: Tuning helps control the amount of oscillation or overshoot in the system response. Excessive oscillations or overshoots can lead to mechanical stress and reduced system lifespan.*
   - *Settling Time: Proper tuning reduces the time it takes for the system to settle to a steady state after a disturbance, improving overall system behavior.*
5. *Optimizing Control System Performance*
   - *Transient Response: Tuning improves the system's transient response, such as the rise time, peak time, and settling time, ensuring it meets the specific requirements of the application.*
   - *Frequency Response: For systems analyzed in the frequency domain, tuning helps achieve the desired gain and phase margins, ensuring robust performance across a range of frequencies.*

6. *Facilitating Design and Implementation*
    - *Design Goals: Proper tuning helps in meeting the design specifications and performance goals set during the control system design phase.*
    - *Implementation Flexibility: Tuning allows for adjustments based on practical implementation constraints and real-world conditions, which might not be fully captured during the design phase.*
7. *Ensuring Safety*
    - *Avoiding Unsafe Conditions: Proper tuning prevents the system from operating in unsafe conditions, reducing the risk of accidents or damage to equipment.*
    - *Compliance: Well-tuned controllers can help ensure that the system complies with safety and performance standards required in various industries.*

Question 2: How does one effectively tune a controller when the system includes a time delay, and what are the key considerations for achieving optimal performance in such systems?

Answer: *Tuning a controller in the presence of a time delay can be challenging due to the impact of the delay on system stability and performance. Here are the key considerations and steps for effectively tuning such a system:*

Understand the Delay Effect

- *Impact on Stability: Time delays introduce phase lag, which can reduce the stability margins of the control system. This phase lag increases with frequency and can lead to oscillations or instability if not properly managed.*
- *Performance Degradation: Delays can affect the speed of response and accuracy of the control system, making it crucial to compensate for the delay to meet performance requirements*

Use of Compensators

- *Lead Compensator: A lead compensator can be used to add phase lead, which helps offset the phase lag introduced by the delay. This can improve stability and transient response.*
- *Smith Predictor: A Smith Predictor is a control strategy specifically designed to handle time delays. It uses a model of the system to predict the effect of the delay and compensate for it in real-time*

Adjust Controller Parameters

- *Proportional-Derivative-Integrative (PID) Tuning: For PID controllers, the tuning parameters (Kp, Ki, Kd) need to be adjusted considering the delay. Typically, the derivative action (Kd) should be carefully tuned to handle the effects of delay without introducing excessive noise.*

- *Frequency Domain Methods: Techniques such as Bode plot analysis can help in adjusting the controller parameters by considering the delay's impact on the phase margin and gain margin.*

Simulation and Testing

- *Simulate the System: Use simulation tools to model the system with the delay included. This allows you to test different tuning parameters and compensators before applying them to the real system.*
- *Test Under Real Conditions: Once a set of parameters is chosen, test the system under real operating conditions to ensure that the tuning meets performance and stability requirements. Robust Control Design*

Question 3: A process reaction curve is a graphical representation used in control engineering to understand the dynamic behavior of a system in response to changes in input. It shows how a system reacts over time to a step change in input, typically from a controller or external disturbance.

Answer:

1. *Step Change: The process reaction curve starts with a sudden change in the input to the system, often referred to as a step input. This could be a sudden change in the control signal, such as opening a valve or changing the setpoint in a process.*
2. *System Response: After the step change, the system's output response is monitored over time. This could be temperature, pressure, flow rate, or any other measurable process variable.*
3. *Curve Plotting: The output of the system is plotted against time to create the reaction curve. This curve typically shows how the output starts from an initial value, changes over time, and eventually settles into a new steady-state value.*
4. *Analysis: By analyzing the process reaction curve, engineers can determine important characteristics of the system, such as the time constant (how quickly the system responds), the gain (how much the output changes for a given change in input), and the delay time (how long it takes for the system to start responding).*
5. *Applications: The process reaction curve is fundamental in designing and tuning control systems. It helps in understanding the system dynamics and aids in selecting appropriate control strategies and parameters.*

In summary, a process reaction curve provides valuable insights into how a system responds to changes, which is crucial for the effective control and optimization of industrial processes.

What is the process reaction curve and how it is used for controller tuning?

The process reaction curve is a key tool in process control and automation, used to analyze and tune control systems. Here's a detailed look at what it is and how it's used for controller tuning:

What is a Process Reaction Curve?

A process reaction curve, also known as a process step response curve, is a graphical representation that shows how a process variable (output) responds over time to a sudden change in input (step change). The step change is typically a sudden adjustment made to the system, such as a change in the input signal from a controller or an external disturbance.

Components of a Process Reaction Curve

1. *Step Change: The input to the system is changed abruptly, usually from one steady-state value to another.*
2. *Response Measurement: The output of the system is measured over time following the step change.*
3. *Plotting the Curve: The output is plotted against time to create the reaction curve, which visually represents the system's dynamic behavior.*

Key Characteristics Revealed by the Curve

1. *Delay Time (L): The time between the step change in input and the initial response of the output. This represents how long it takes for the system to start reacting.*
2. *Rise Time (T_r): The time taken for the output to rise from the initial value to a significant portion of its final value (often 90% of the final value).*
3. *Time Constant (T): The time required for the output to reach approximately 63.2% of its final steady-state value after the initial response. This is indicative of how quickly the system responds.*
4. *Steady-State Gain (K): The ratio of the change in output to the change in input after the system has reached a new steady state. This reflects how much the output changes in response to a change in input.*
5. *Settling Time: The time required for the output to remain within a certain percentage of the final value (e.g., 5% or 2%).*

Using the Process Reaction Curve for Controller Tuning

The process reaction curve is essential for tuning controllers, particularly in methods like the Ziegler-Nichols tuning method. Here's how it helps in controller tuning:

1. *Determine System Parameters: By analyzing the process reaction curve, you can determine the key system parameters (delay time, time constant, and steady-state gain). These parameters are crucial for designing and tuning controllers.*
2. *Choose Controller Type: Based on the system characteristics, you can decide whether to use a Proportional (P), Proportional-Integral (PI), or Proportional-Integral-Derivative (PID) controller.*
3. *Apply Tuning Rules: Tuning methods such as the Ziegler-Nichols method use the process reaction curve parameters to set the initial values for the controller gains. For instance:*

- *Ziegler-Nichols Open-Loop Method: Use the delay time (L) and time constant (T) from the reaction curve to calculate the proportional gain (K_p), integral time (T_i), and derivative time (T_d).*
- *Ziegler-Nichols Closed-Loop Method: Perform an open-loop test and then adjust the controller gains based on the system's oscillatory behavior to achieve the desired performance.*

4. *Optimize Performance: With the initial parameters set, you can fine-tune the controller to achieve optimal performance, such as minimizing overshoot, reducing settling time, and ensuring stability.*

Question: 4 For feedback controller tuning purposes, the most commonly employed approximate process model form is the First-Order Plus Dead Time (FOPDT) model.

Answer: *This model is widely used due to its simplicity and effectiveness in representing a broad range of real-world processes.*

Key Components of the FOPDT Model:

1. *Steady-State Gain (K): This parameter indicates how much the output of the process will change in response to a unit change in the input. It represents the ratio of the output change to the input change when the system is in a steady state.*
2. *Time Constant (τ): This parameter reflects the speed of the process's response to input changes. It describes how quickly the process moves towards its new steady state after a disturbance.*
3. *Dead Time (θ): This represents the delay between the input change and the beginning of the process's response. It accounts for the time it takes for the effect of the input change to become visible in the process output.*

Why FOPDT is Popular for Controller Tuning

1. *Simplicity: The FOPDT model is relatively simple and provides a good balance between model complexity and accuracy. It captures the essential dynamics of many processes without requiring complex modeling.*
2. *Tuning Rules: Many controller tuning methods, such as the Ziegler-Nichols method, the Cohen-Coon method, and others, are based on the FOPDT model. These methods use the parameters obtained from the FOPDT model to set initial values for controller gains.*
3. *Practicality: The FOPDT model is practical for real-world applications where processes have some degree of dead time and first-order dynamics. It is often sufficient for designing and tuning PID controllers.*

Application in Controller Tuning

In controller tuning, the FOPDT model helps engineers determine appropriate values for proportional, integral, and derivative gains by providing a straightforward way to estimate system dynamics. For example, using the FOPDT parameters, one can apply tuning rules to set the PID parameters and achieve desired performance characteristics, such as stability, response time, and minimal overshoot.

The fundamental principle behind using stability margins for controller design when process models are available is to ensure that the closed-loop control system remains stable and performs adequately even in the presence of uncertainties and variations in the process dynamics.

What is the fundamental principle behind the use of stability margins for controller design when process models are available?

Key Concepts in Stability Margins:

1. *Stability Margins: Stability margins are measures that indicate how close a control system is to becoming unstable. They are used to assess and ensure robustness in the control design. The primary stability margins are:*
   - *Gain Margin (GM): The amount by which the system gain can be increased before the system becomes unstable. It is the reciprocal of the gain at the phase crossover frequency (where the phase of the open-loop transfer function is -180 degrees).*
   - *Phase Margin (PM): The amount by which the phase of the system can be decreased before it becomes unstable. It is the difference between the phase of the open-loop transfer function at the gain crossover frequency (where the gain is 1 or 0 dB) and* $-180°$.
2. *Frequency Domain Analysis: Stability margins are typically analyzed using frequency domain methods, such as Bode plots and Nyquist plots. These methods provide insights into the behavior of the system across different frequencies and help in determining the stability margins.*
3. *Robustness: Stability margins are crucial for robustness. They ensure that the system will remain stable and perform well even when there are changes in process dynamics, parameter variations, or unmodeled dynamics. Robust controllers are designed to handle such uncertainties effectively.*

Application in Controller Design:

1. *Model-Based Design: When process models are available, the design of controllers can be based on these models. The models provide information about the process dynamics, which can be used to predict the closed-loop system behavior and analyze stability margins.*
2. *Tuning Controllers: Stability margins help in tuning controllers to achieve desired performance while maintaining stability. By analyzing the frequency response of the system, engineers can adjust the controller parameters to meet specific gain and phase margin requirements.*
3. *Design Criteria: Stability margins serve as design criteria to ensure that the controller design is robust. For instance:*

- *Gain Margin: A higher gain margin implies that the system can tolerate a larger increase in gain before becoming unstable. This provides a buffer against unexpected changes or disturbances.*
- *Phase Margin: A higher phase margin indicates a more stable system, with a larger tolerance for phase lag due to modeling errors or additional dynamics.*
- *Performance and Stability Trade-offs: Stability margins also help balance performance and stability. For example, increasing the phase margin generally improves stability but may slow down the response. Engineers use stability margins to find an optimal balance that meets both performance and stability requirements*

4. *Performance and Stability Trade-offs: Stability margins also help balance performance and stability. For example, increasing the phase margin generally improves stability but may slow down the response. Engineers use stability margins to find an optimal balance that meets both performance and stability requirements.*

References

1. Stephanopoulos G (1984) Chemical process control. Englewood Cliffs, New Jersey, Prentice Hall
2. Nagarsheth SH, Sharma SN (2020) Control of non-minimum phase system with dead time: a fractional system viewpoint. Int J Syst Sci 51(11):1905–1928
3. Ranganayakulu R, Rao AS, Bhasakar B (2020) Analytical design of fractional IMC filter–PID control strategy for performance enhancement of cascade control systems. Int J Syst Sci 51(10):1699–1713
4. Raja GL, Ali A (2021) Enhanced tuning of Smith predictor-based series cascade control structure for integrating. ISA Trans 114:191–205
5. Yumuk E, Güzelkaya M, Eskin I (2019) Analytical fractional PID controller design based on Bode's ideal transfer function plus time delay. ISA Trans 91:196–206
6. Khalifa B, Abdelfateh C (2017) Optimal tuning of fractional order a controller using particle swarm optimization algorithm. IFAC-OnLine 50(1):8084–8089
7. Begum KG, Rao AS, Radhakrishnala TK (2017) Enhanced IMC-based PID controller design for non-minimum phase (NMP) integrating processes with time delays. ISA Trans 68:223–234
8. Pai NS, Chang SC, Huang CT (2010) Tuning PI/PID controllers for integrating processes with dead-time and inverse response by simple calculations. J Process Control 20(6):726–733
9. Jeng JC, Lin SW (2012) Robust proportional-integral derivative controller design for stable/integrating processes with an inverse response and time delay. Ind Eng Chem Res 51(6):2652–2665
10. Joshi D, Deb D, Giri AK (2024) MRAS disturbance observer-based sensorless field-oriented backstepping control of BLDC motor drive. In: Electrical engineering. Springer, Berlin. https://doi.org/10.1007/s00202-024-02378
11. Kundu S, Singh M, Giri AK (2022) Implementation of in-loop filter based DSC-FLL control approach for power quality improvement in isolated microgrid. In: Proceedings of 2022 IEEE 31st international symposium on industrial electronics (ISIE). USA. https://doi.org/10.1109/ISIE51582.2022.9831712
12. Giri AK, Arya SR, Maurya R (2020) Hybrid order generalized integrator based control for VSC to improve the PMSG operation in isolated mode. In: The proceedings of 2020 first international

conference on power, control and computing technologies (ICPC2T). India. https://doi.org/10.1109/ICPC2T48082.2020.9071498
13. Kundu S, Singh M, Giri AK (2024) Synchronization and control of WECS-SPV-BSS-based distributed generation system using ICCF-PLL control approach. Electr Power Syst Res 226:109919
14. Giri AK, Arya SR, Guerrero JM, Kumar S, Power quality: infrastructures and control. Springer Nature, Singapore. ISBN: 978-981-19-7956-9
15. Giri AK, Arya SR, Vinnikov D, Distributed energy systems design, modeling, and control, vol 1. CRC Press, pp 1–308. ISBN: 9781032134246
16. Arya SR, Patel MM, Alam SJ, Srikakolapu J, Giri AK, Chitti Babu B (2021) Classical control algorithms for permanent magnet synchronous generator driven by diesel engine for power quality. Int J Circuit Theory Appl 49(3):576–601
17. Kundu S, Giri AK, Singh M, Kadiyan S, Adaptive control strategy for isolated renewable energy-based generation system with intermittency. In: Electrical engineering. Springer, Berlin. https://doi.org/10.1007/s00202-024-02454-0
18. Meena DC, Singh M, Giri AK (2021) Leaky-momentum control algorithm for voltage and frequency control of three-phase SEIG feeding isolated load. J Eng Res ICARI Spec 109:120
19. Meena DC, Singh M, Giri AK (2021) A modified NLMS control algorithm for coordinated operation in three-phase wind-energy conversion system. Int J Renew Energy Res (IJRER) 11(4):1621–1629
20. Dalwadi N, Deb D, Ozana S (2023) Performance evaluation of RW-quadrotor and Bi-quadrotor for payload delivery. IET Intel Transp Syst 17(11):2221–2236
21. Dalwadi N, Deb D, Ozana S, Adaptive hybrid control of quadrotor drones. 461:1–174. ISBN: 978-981-19-9743-3

Chapter 6
Tuning of Fractional Order Filter with Non-minimum Phase Zero

6.1 An Improved Series Cascade Control Structure

Section 6.1 outlines the Bode ideal transfer function, which incorporates a non-minimum phase (NMP) zero.

$$B'(s) = \frac{(1 - as)}{\phi s^{\lambda}}. \tag{6.1}$$

In [1], Bode's ideal transfer function guided the design of a state feedback controller with integral action [2], on the other hand, proposed a robust fractional-order controller within the IMC framework, employing Bode's ideal transfer function for tuning the fractional filter in stable higher-order plants. This chapter unveils a novel tuning strategy aimed at non-minimum phase systems and fractional filter adjustment using Bode's ideal transfer function, particularly considering NMP zeros [3].

Bodes' Ideal transfer function

The open-loop representation of Bode's ideal transfer function takes the form, [2]

$$B'(s) = \frac{1}{\phi s^{\lambda}}, \lambda \in R. \tag{6.2}$$

The tuning expression for the fractional filter parameters is specified [2]

$$\lambda = \frac{\pi - P.M.}{\pi/2} - 1, \phi = (\omega)^{-\lambda}.$$

M. Yadav et al., *Control Strategies for Non-minimum Phase Systems*,
Studies in Infrastructure and Control, https://doi.org/10.1007/978-981-96-2351-8_6

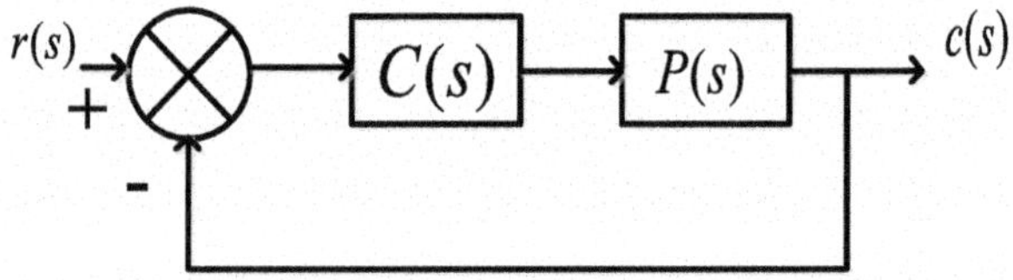

Fig. 6.1 Traditional control system

Suggested tuning expressions

The parameters of the fractional filter in this proposed setup are tuned using Bode's ideal transfer function. After this adaptation, the expression for tuning the fractional filter is detailed as follows:

$$\lambda = \frac{\frac{\pi}{2} - PM}{\frac{\pi}{2}}, \phi = \frac{\sqrt{1 + \omega^2 b^2}}{(\omega)^{\delta+1}}.$$

From Fig. 6.1,

$$C(s) = H(s)K_c\left(1 + \frac{1}{T_I s} + T_D s\right),$$

Here $H_r(s)$ lies the fractional filter, while the rest constitutes the PID controller. The following are the three conditions that the controller must meet.

Situation 1: Conditions governing the phase margin: $Arg(C(j\omega_c)P(j\omega_c)) = -\pi + PM$.

Situation 2: Specifications for the gain cross-over frequency. $|C(j\omega_c)P(j\omega_c)| = 1$.

Situation 3: Nullification of the steady-state error.

Procedure for adaption

1. Equation (6.1), upon substitution $s = j\omega$, becomes,

$$B'(j\omega) = \frac{(1 - aj\omega)}{\phi(j\omega)^{\lambda+1}}. \tag{6.3}$$

2. Upon adhering to situation 1 and performing the argument operation on both sides of equation (6.3), the expression can be stated as,

$$2\pi - \frac{\pi}{2} - \frac{\lambda\pi}{2} = \pi + PM. \tag{6.4}$$

3. Subsequently simplifying Eq. (6.4)

$$\lambda = \frac{\frac{\pi}{2} - PM}{\frac{\pi}{2}}. \tag{6.5}$$

4. Moreover, when the modulus of Eq. (6.5) is considered,

$$|B'(j\omega)| = \frac{\sqrt{1+\omega^2 a^2}}{\phi(\omega)^{\lambda+1}}. \tag{6.6}$$

5. Now resulting condition (2) then Eq. (6.6) can be written as,

$$1 = \frac{\sqrt{1+\omega^2 a^2}}{\phi(\omega)^{\lambda+1}}. \tag{6.7}$$

6. Subsequently, in the simplification Eq. (6.7), the time constant is realized as,

$$\phi = \frac{\sqrt{1+\omega^2 a^2}}{(\omega)^{\lambda+1}}. \tag{6.8}$$

6.2 Controller Design

Here we sketched and tuned the controller after the utilization of a novel tuning methodology,

Inner loop controller design: Refer to Sect. 2.10 for details on the inner loop controller

$$C_2(s) = \frac{T_2 s + 1}{K_2'(\eta s + 1)}. \tag{6.9}$$

Outer loop controller design: In the context of designing an outer loop controller using the Internal Model Control (IMC) approach, and facing an integrating process with a non-minimum phase (NMP) zero, several different methods can be considered.

$$\tilde{P}_1(s) = \frac{K_1'(1-as)}{s}.$$

The plant's complete transfer function is expressed as,

$$\tilde{P}(s) = H(s)\tilde{P}_1(s). \tag{6.10}$$

The term $H(s)$ can be written as,

$$H_r(s) = \frac{C_2(s)P_2(s)}{1 + C_2(s)(P_2(s) - \tilde{P}_2(s))}.$$

From Eq. (6.6),

$$\tilde{P}(s) = \frac{K(1-as)}{s(\eta s+1)}. \tag{6.11}$$

Leveraging the equivalent system concepts from Figs. 6.2 and 6.3, the transfer function for the controller can be calculated.

$$B'(s) = \frac{C_1(s)P(s)}{1+C_1(s)I_o(s)}. \tag{6.12}$$

Following the modification of Eq. (6.12), the universal form of the proposed controller (6.13) evolves to.

In this context, the inverse response compensator represents the signal that the controller

$$C_1(s) = \frac{B'(s)}{\tilde{P}(s)[1-B'(s)I_o(s)]}. \tag{6.13}$$

receives, implying it is from a typical system. Once Eqs. (6.11) and (6.12) are substituted into (6.13), it can be adapted as

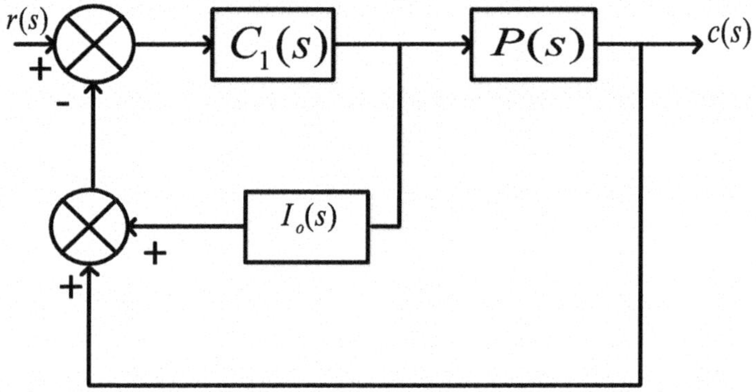

Fig. 6.2 Feedback controller with inverse response compensator

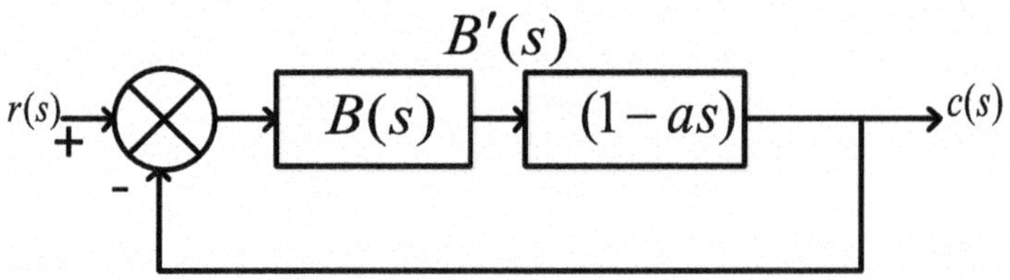

Fig. 6.3 Non-minimum phase version of Bode's ideal transfer function

$$C_1(s) = \frac{s^{1-\lambda}}{\left[\phi + 2a^2 s^{2-\lambda} - 2as^{1-\lambda}\right]}. \tag{6.14}$$

6.3 Search Space Reduction (SSR) Algorithm

This chapter unveils the novel metaheuristic algorithm, the Search Space Algorithm (SSR), as illustrated in [5]. In contrast to conventional methods that update solutions iteratively, SSR randomly adjusts solutions within a diminished search space during each iteration [6].

Concluding the initial search boundary: Eq. (6.15) elucidates the upper and lower limits. $t' = 1$.

$$m(t') = \frac{ub - lb}{2}. \tag{6.15}$$

In Eq. (6.15), it is noted that search agents remain within predefined limits. If Eqs. (6.16) result in search agents moving beyond the original bounds, it can be expressed as follows:

$$X_i'(t') = lb + (lb \cdot ub) \times rand_o \text{ if } X_i' < lb \text{ or } X_i' > ub \tag{6.16}$$

Computing the middle position: Following the application of Eq. (6.17), the midpoint evaluation can be carried out with respect to the best solution M.

$$mid_o(t') = mean(X'_{bestM}). \tag{6.17}$$

If the user decides on $M = 2$ This selection will determine the three universal paramount positions X'_{gbest1}, X'_{gbest2} respectively.

$$mid_o(t') = \frac{X'_{best1} + X'_{best2}}{2}$$

Adjusting the exploration scope: As per Eq. (6.18), the search range decreases with each iteration

$$m'(t) = m'(t' - 1) \times e^{(R_f / t'_{\max})}. \tag{6.18}$$

Restructuring of search agents: The ensuing iteration is reconfigured close to the midpoint with the aid of an updated search agent. This can be articulated as:

$$X_i'(t') = mid'(t') + rand' \times m'(t'') \quad \text{if}\, rand' >= 0.5.$$

$$X_i'(t') = mid'(t) - rand' \times m'(t') \quad \text{if}\, rand' <= 0.5.$$

Figure 6.4 displays the flowchart of the SSR algorithm, while Fig. 6.5 outlines the recommended process for tuning controllers using the SSR algorithm, directly indicating PI and PID controller parameters within the system parameters.

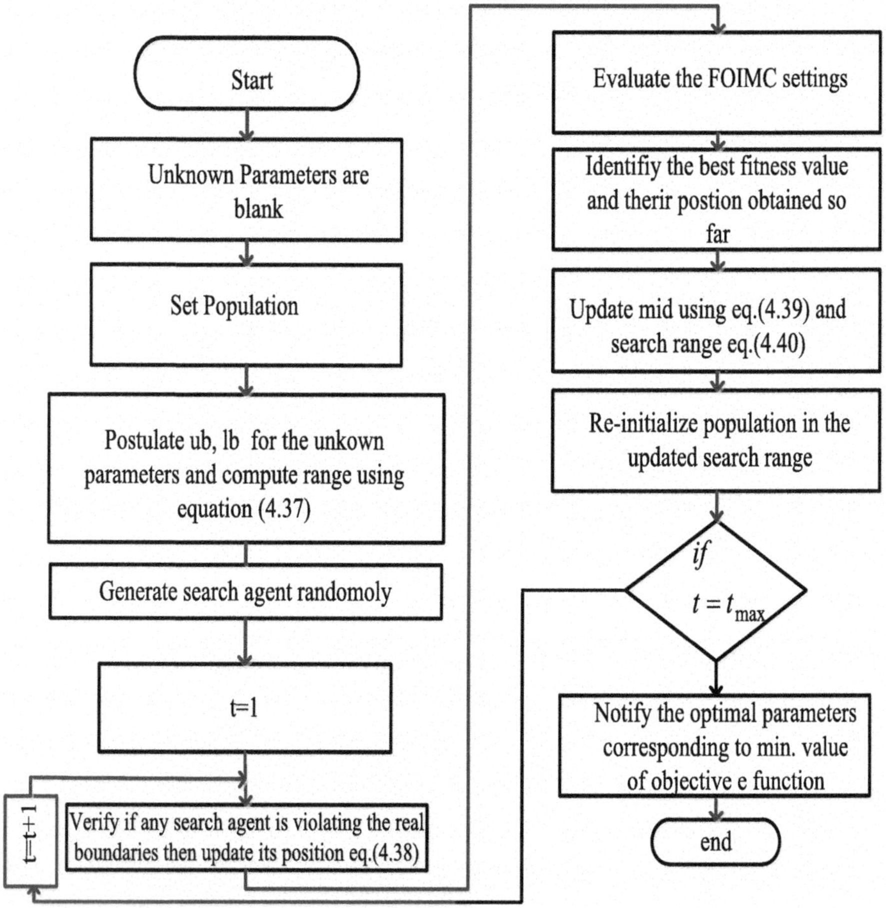

Fig. 6.4 Structural outline of SSR algorithm

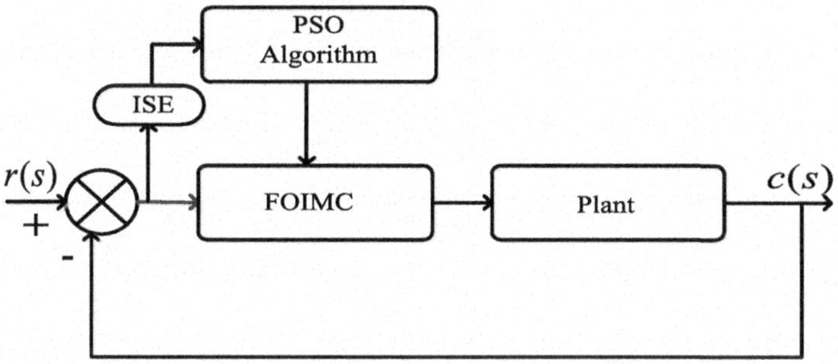

Fig. 6.5 FOIMC controller with SSR algorithm

6.4 Case Studies

The effectiveness of the recommended strategies is evaluated through the analysis of two case studies, assessing their performance under nominal conditions and with a +10% model mismatch.

6.4.1 *Case Study I*

The Endorsement is given to the transfer function of the higher-order system configuration [2],

$$\tilde{P}_1(s) = \frac{(1 - 0.4699s)}{s}, \tilde{P}_2(s) = \frac{0.5183}{1.1609s + 1}.$$

Inner loop controller design: From Eq. (6.9)

$$C_2(s) = \frac{1.1609s + 1}{0.5183(4s + 1)}.$$

With Eq. (6.11), the overall plant transfer function can be expressed as

$$\tilde{P}(s) = \frac{(1 - 0.4699s)}{s(4s + 1)}. \tag{6.19}$$

Outer loop controller design: From Eq. (6.14)

$$C_1(s) = \frac{s^{0.06}}{1.1 + 0.44s^{1.06} - 0.93s^{0.06}}\left(1 + \frac{0.25}{s}\right), \tag{6.20}$$

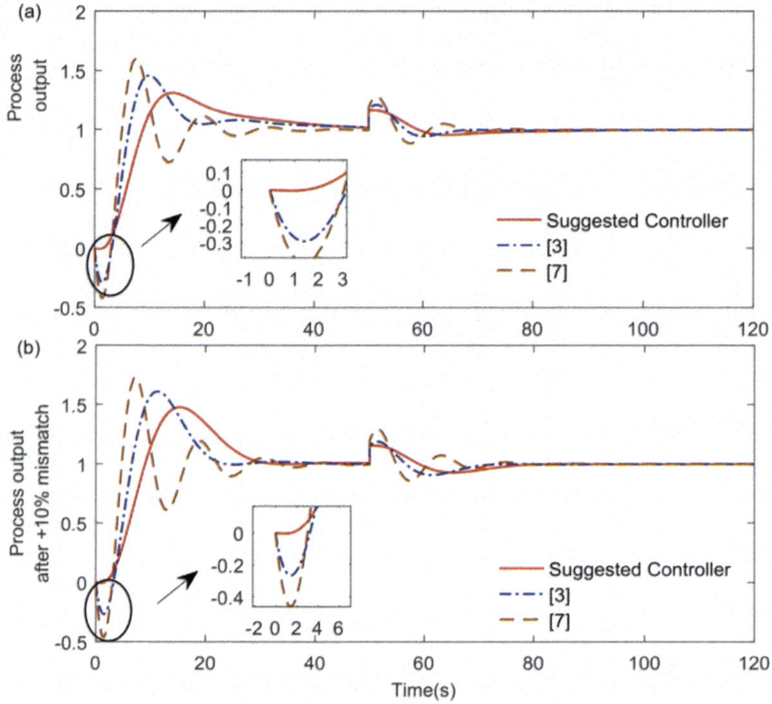

Fig. 6.6 Closed-loop response of higher-order integrating process

where the fractional filter is $\frac{s^{0.06}}{1.1+0.44s^{1.06}-0.93s^{0.06}}$ with $\phi = 1.1$, $\lambda = 0.93$, $\eta = 4$, and $PM = 65$. The comparative analysis in Fig. 6.6 evaluates the proposed controller alongside two alternative control structures, with all control loops encountering a disturbance of 0.15 at t = 50 for both proposed and existing configurations.

A comprehensive closed-loop depiction of the fractional characteristic polynomial for the system is outlined in the report.

$$f(s) = 1 + 1.1s^{0.94} + 0.44s^2 - 0.93s.$$

The stability examination is visually presented in Fig. 6.7. Table 6.1 provides a list of the controller performance indices and transient responses.

The sensitivity function is denoted by [2],

$$S(s) = \frac{1}{1 + \tilde{P}(s)C_p(s)}. \tag{6.21}$$

To illustrate robustness, substitute the plant and controller transfer function values from Eqs. (6.19) and (6.20) into (6.21). Figure 6.8a, b show the graphical representation of the sensitivity function for the proposed controller compared with the other

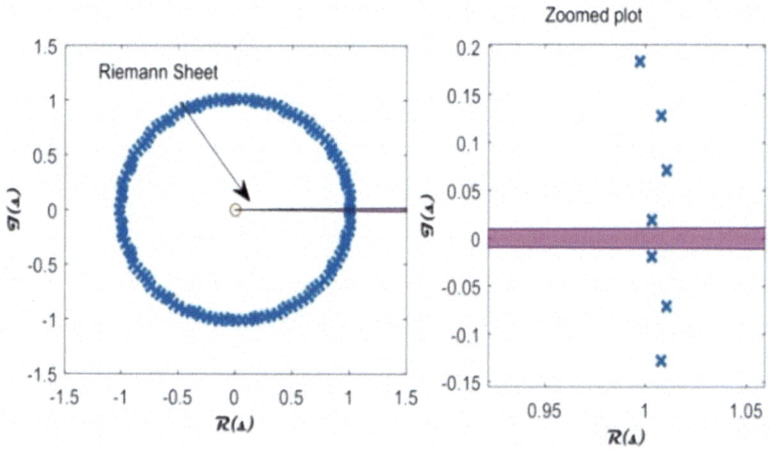

Fig. 6.7 Stability investigation of higher-order integrating process

Table 6.1 Performance assessment and transient outcome for higher order integrating process

Method	Nominal case				+10% Perturbed case			
	Peak O.S. ratio, %	Peak time, (s)	IAE	ISE	Peak O.S. ratio, %	Peak time, (s)	IAE	ISE
Suggested controller	1.3	15.47	0.011	0.000130	1.48	14.43	0.10	0.010
Begum et al. [3]	1.4	11.21	0.15	0.024	1.61	9.9	0.97	0.95
Kaya [7]	1.60	7.5	2.4	6.1	1.72	7.3	2.6	7.1

controllers. The effectiveness of the SSR algorithm is highlighted in Fig. 6.9, which shows the convergence of the objective function with the optimal location adjusted slightly from its original position.

The effectiveness of the SSR algorithm is evident in Fig. 6.10a, b, where the proposed method exhibits considerable performance improvements in both nominal and +10% mismatch cases following optimization.

Evaluation outcomes show that the suggested algorithm offers practical applicability and robust performance.

Refer to Table 6.2 for detailed parameter descriptions.

Remark 6.1 A comparison is made between the suggested method and those described in [3, 7], with all parameters held constant. The approaches in [3, 7] used higher-order model approximations and setpoint filters but lacked an essential inverse response compensator and did not utilize an optimization algorithm for optimizing controller parameters.

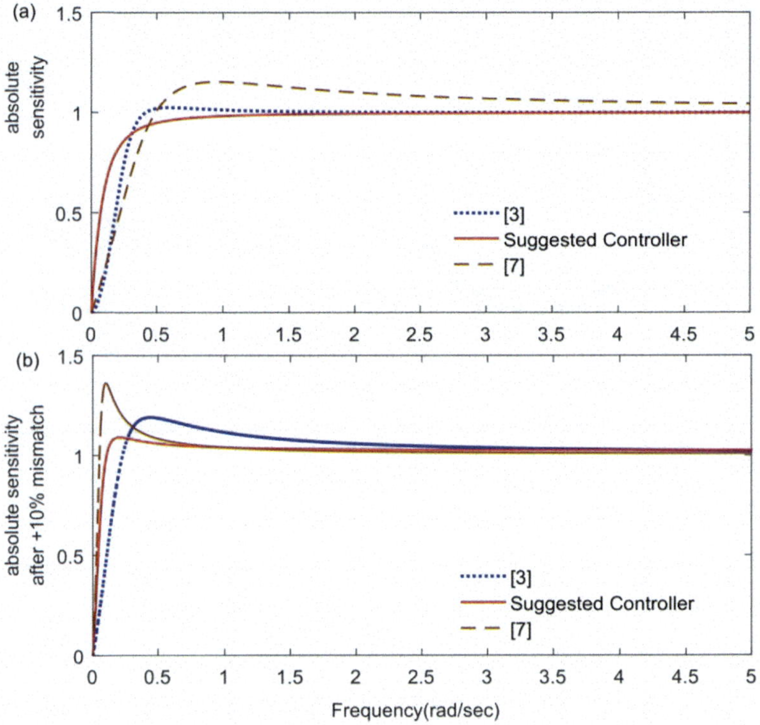

Fig. 6.8 Robustness analysis of the suggested method

Fig. 6.9 Convergence curve of the objective function

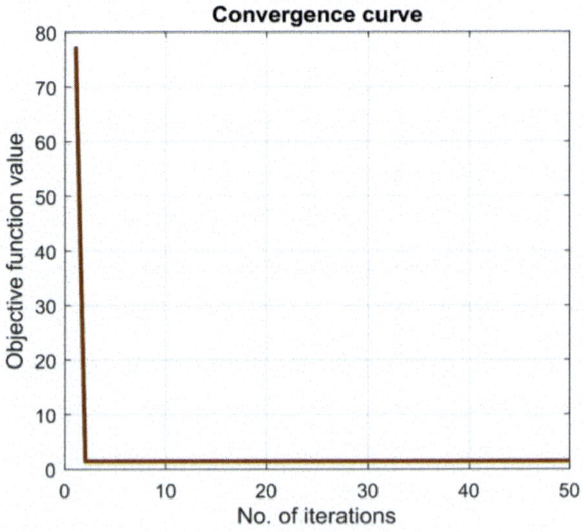

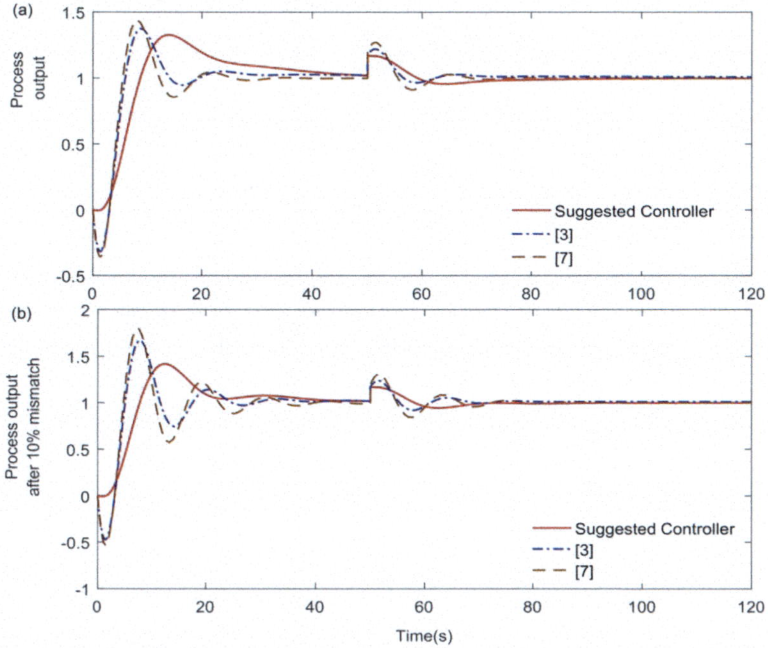

Fig. 6.10 Closed-loop response via SSR algorithm

Table 6.2 Parameters of the SSR algorithm

Parameter	No. of iterations	M	R_f	Search agent
Value	50	2	15	25

6.4.2 *Case Study II*

The transfer function representing the boiler drum level system has been validated [2]

$$\tilde{P}_1(s) = \frac{(1 - 0.418s)}{s}, \tilde{P}_2(s) = \frac{0.547}{1.06s + 1}.$$

Inner loop controller design: From Eq. (6.9)

$$C_2(s) = \frac{1.060s + 1}{0.547(3.5s + 1)}.$$

Now using Eq. (6.11), we can determine the overall plant transfer function.

$$\tilde{P}(s) = \frac{(1 - 0.418s)}{s(3.5s + 1)}. \tag{6.22}$$

Outer loop controller design: From Eq. (6.14) the outer loop controller becomes,

$$C_1(s) = \frac{s^{0.23}}{1.2 + 0.34s^{1.23} - 0.83s^{0.23}}\left(1 + \frac{0.28}{s}\right), \tag{6.23}$$

where the fractional filter is $\frac{s^{0.23}}{1.2+0.34s^{1.23}-0.83s^{0.23}}$ with $\lambda = 0.77, \phi = 1.2$, $\eta = 3.5$ and $PM = 72$. Figure 6.11a, b illustrates the step response in both the cases (nominal and +10% mismatch) of the boiler drum level system with an amplitude of 0.15 at t = 20 s.

The stability analysis of the boiler drum level system is detailed in this section,

$$f(s) = 1 + 1.2s^{0.77} + 0.34s^2 - 0.83s$$

Select $\alpha = s^{\beta}$ The associated characteristic polynomial is,

$$f(\alpha^{1/\beta}) = 1 + 1.2\alpha^{770} + 0.34\alpha^{2000} - 0.83\alpha^{1000}.$$

Where, $s^{0.001} = \alpha$. Figure 6.12 shows the roots outside the Riemann sheet, which provides a demonstration of the system's stability.

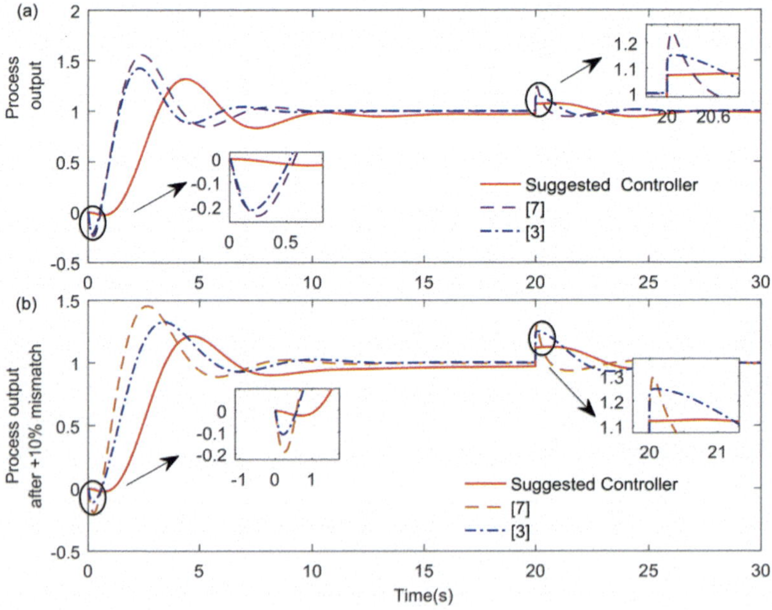

Fig. 6.11 Relative investigation of the suggested approach

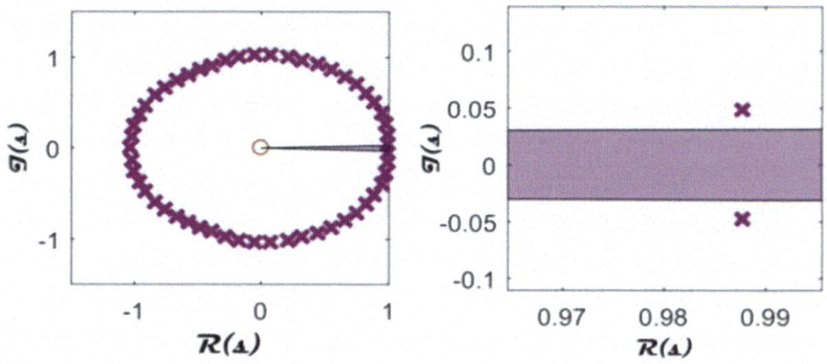

Fig. 6.12 Stability investigation of boiler drum level system

The greater details of robustness analysis are presented in case study 6.1. In Fig. 6.13, sensitivity analyses are depicted for both nominal and +10% mismatch scenarios. Figure 6.13 interprets the suggested controller has a minimum value of sensitivity function which can indicate better robustness.

In Fig. 6.14 the graph depicting the convergence of the Integral Squared Error (ISE) objective function is presented.

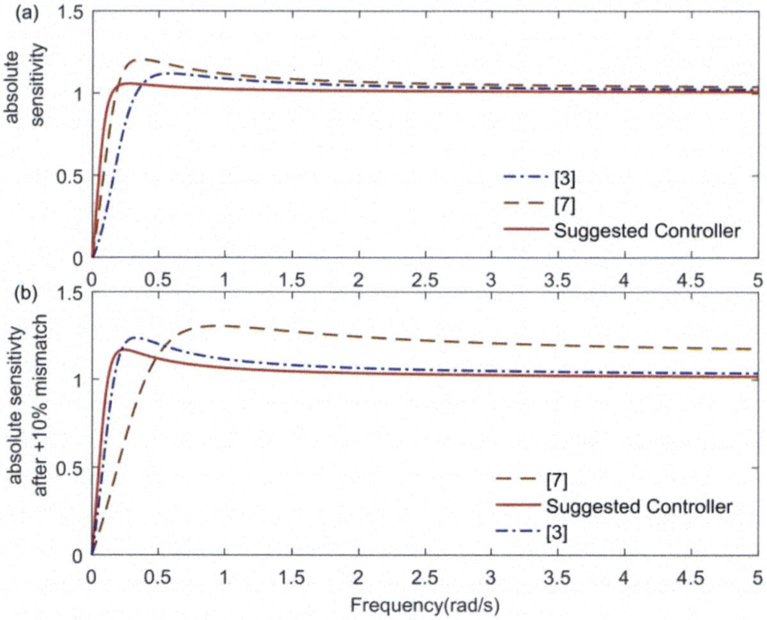

Fig. 6.13 Robustness analysis of the proposed method

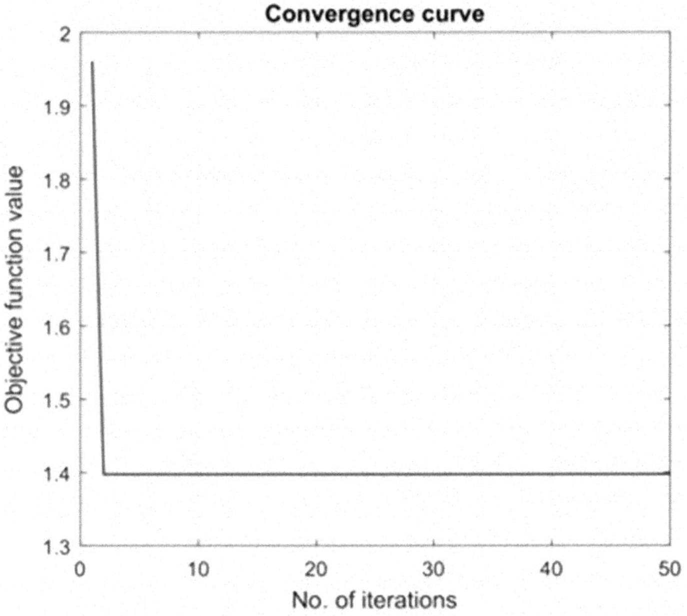

Fig. 6.14 Convergence graph of ISE objective function

The application of SSR results in Fig. 6.15 shows an improved closed-loop response for the proposed controller. Furthermore, Table 6.3 reports decreases in Integral Absolute Error (IAE), Integral Squared Error (ISE), and overshoot.

Remark 6.2 The effectiveness of the proposed method was evaluated through a comparative study with methodologies presented in [3, 7]. In [7], overshoot compensation was achieved using a setpoint filter, but controller settings were not optimized. Similarly, [3] did not optimize controller settings or include an inverse response compensator.

6.5 Summary

This chapter refines Bode's ideal transfer function to account for non-minimum phase (NMP) zeros and introduces a unique SSR algorithm-based tuning method for non-minimum phase systems. This method allows for the integrated management of dead time and NMP zeros, providing the foundation for the subsequent discussions in Chap. 7.

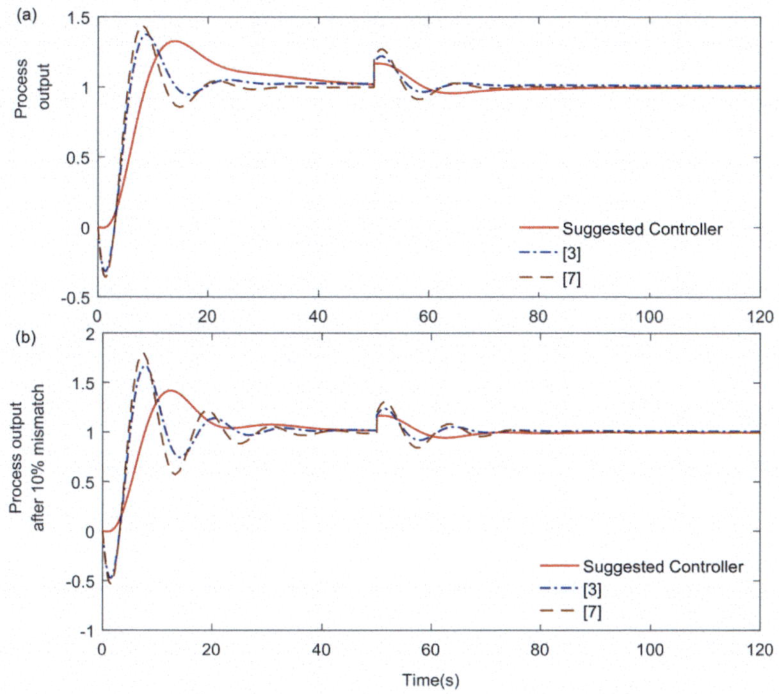

Fig. 6.15 Comparative analysis of SSR algorithm

Table 6.3 Performance assessment and transient outcome for boiler drum level system

Method	Nominal case				+10% Perturbed case			
	Peak O.S. ratio, %	Peak time, (s)	IAE	ISE	Peak O.S. ratio, %	Peak times, (s)	IAE	ISE
Suggested controller	1.25	7.4	0.14	0.019	1.48	5.9	0.19	0.037
Begum et al. [3]	1.32	3.4	0.28	0.082	1.61	2.4	0.34	0.12
Kaya [7]	1.45	2.8	0.43	0.18	1.72	2.3	0.49	0.24

6.6 Some Informative Views of Chapter in Post-mortem Form

Question 1: Define the desired frequency response characteristics of the filter.

Answer: *The cutoff frequency, bandwidth, and attenuation levels are the desired frequency response characteristics. The fractional order affects these parameters differently compared to integer-order filters.*

Question 2: Why is tuning of a controller important in control system design?

Answer: *The primary reason for tuning a controller is to achieve optimal performance of the control system. Controllers, such as PID (Proportional-Integral-Derivative), adjust outputs based on feedback to regulate processes or systems. Proper tuning ensures that the controller responds efficiently to changes and disturbances, achieving desired setpoints quickly and accurately.*

Question 3: How does tuning a controller enhance efficiency?

Answer: *The controller tuning optimizes control system performance across industries by improving response times, managing energy use, ensuring stability, adapting to changes, and reducing operational expenses. These benefits collectively contribute to sustainable and cost-effective operations, making tuning a critical aspect of control system design and operation.*

Question 4: What are the advantages of an effective tuning approach?

Answer: *The effective tuning approach has the following benefits.*

Improved Performance

Enhanced Stability

Reduced Energy Consumption

Increased Productivity

Adaptability

Extended Equipment Lifespan

Improved Quality

References

1. Raja GL, Ali A (2021) Enhanced tuning of Smith predictor-based series cascade control structure for integrating. ISA Trans 114:191–205
2. Nagarsheth SH, Sharma SN (2020) Control of non-minimum phase system with dead time: a fractional system viewpoint. Int J Syst Sci 51(11):1905–1928
3. Begum KG, Rao AS, Radhakrishnala TK (2017) Enhanced IMC-based PID controller design for non-minimum phase (NMP) integrating processes with time delays. ISA Trans 68:223–234
4. Saxena S, Hote YV (2022) Design of robust fractional-order controller using the Bode ideal transfer function approach in IMC paradigm. Nonlinear Dyn 107(2):983–1001
5. Aeidapu M, Gangireddy S (2021) A novel search space optimization algorithm. Soft Comput 25(2):9455–9482

6. Lakshmanaprabu SK, Elhoseny M, Shankar K (2016) Optimal tuning of decentralized fractional-order PID controllers for TITO process using the equivalent transfer function. Cogn Syst Res 58:292–303
7. Kaya I (2020) Integral-proportional derivative tuning for optimal closed-loop responses to control integrating processes with an inverse response. Trans Inst Meas Control 42(16):3123–3134
8. Joshi D, Deb D, Giri AK (2024) MRAS disturbance observer-based sensorless field-oriented backstepping control of BLDC motor drive. In: Electrical engineering. Springer, Berlin. https://doi.org/10.1007/s00202-024-02378
9. Kundu S, Singh M, Giri AK (2022) Implementation of in-loop filter based DSC-FLL control approach for power quality improvement in isolated microgrid. In: Proceedings of 2022 IEEE 31st international symposium on industrial electronics (ISIE). USA. https://doi.org/10.1109/ISIE51582.2022.9831712
10. Giri AK, Arya SR, Maurya R (2020) Hybrid order generalized integrator based control for VSC to improve the PMSG operation in isolated mode. In: The proceedings of 2020 first international conference on power, control and computing technologies (ICPC2T). India. https://doi.org/10.1109/ICPC2T48082.2020.9071498
11. Kundu S, Singh M, Giri AK (2024) Synchronization and control of WECS-SPV-BSS-based distributed generation system using ICCF-PLL control approach. Electric Power Syst Res 226:109919
12. Giri AK, Arya SR, Guerrero JM, Kumar S, Power quality: infrastructures and control. Springer Nature, Singapore. ISBN: 978-981-19-7956-9
13. Giri AK, Arya SR, Vinnikov D, Distributed energy systems design, modeling, and control, vol 1. CRC Press, pp 1–308. ISBN: 9781032134246
14. Arya SR, Patel MM, Alam SJ, Srikakolapu J, Giri AK, Chitti Babu B (2021) Classical control algorithms for permanent magnet synchronous generator driven by diesel engine for power quality. Int J Circuit Theory Appl 49(3):576–601
15. Kundu S, Giri AK, Singh M, Kadiyan S, Adaptive control strategy for isolated renewable energy-based generation system with intermittency. In: Electrical engineering. Springer, Berlin. https://doi.org/10.1007/s00202-024-02454-0
16. Meena DC, Singh M, Giri AK (2021) Leaky-momentum control algorithm for voltage and frequency control of three-phase SEIG feeding isolated load. J Eng Res ICARI Spec 109:120
17. Meena DC, Singh M, Giri AK (2021) A modified NLMS control algorithm for coordinated operation in three-phase wind-energy conversion system. Int J Renew Energy Res (IJRER) 11(4):1621–1629
18. Dalwadi N, Deb D, Ozana S (2023) Performance evaluation of RW-quadrotor and Bi-quadrotor for payload delivery. IET Intel Transp Syst 17(11):2221–2236
19. Dalwadi N, Deb D, Ozana S, Adaptive hybrid control of quadrotor drones. 461:1–174. ISBN: 978-981-19-9743-3

Chapter 7
A Combined Approach for Tuning Optimal Fractional IMC-Based Series Cascade Control

7.1 Introduction

PID controllers are widely favored in industry due to their simple tuning methods. The development of fractional calculus-based extensions to conventional integer-order PID controllers is recognized as a significant area for future research in controller design [1].

References [2, 3] suggested a systematic and robust synthesis tuning approach of fractional-order controllers via genetic algorithm. However, after encountering time delay and Non-minimum phase zeros, the controller has additional phase lag and bandwidth. Hence for better closed-loop response effective tuning is essential.

The literature addresses Bode's ideal transfer function with a dubious delay term in optimal algorithm tuning. However, there is a notable gap regarding the incorporation of non-minimum phase zeros and dead time into Bode's ideal transfer function. This chapter illustrates an enhanced model of Bode's ideal transfer function.

7.2 Suggested Work

In this section, both the classic and adapted versions of Bode's ideal transfer function are discussed [4]. Every controller must adhere to these three conditions.

Performance specification 1: Criterion defining phase margin $Arg(C(j\omega)P(j\omega)) = -\pi + PM$.

Performance specification 2: Criteria defining the gain cross-over frequency $|C(j\omega)G(j\omega)| = 1$.

Performance specification 3: Achieving error-free steady-state response.

Also, Fig. 7.1 demonstrates the advised Bode's ideal transfer function,

M. Yadav et al., *Control Strategies for Non-minimum Phase Systems*,
Studies in Infrastructure and Control, https://doi.org/10.1007/978-981-96-2351-8_7

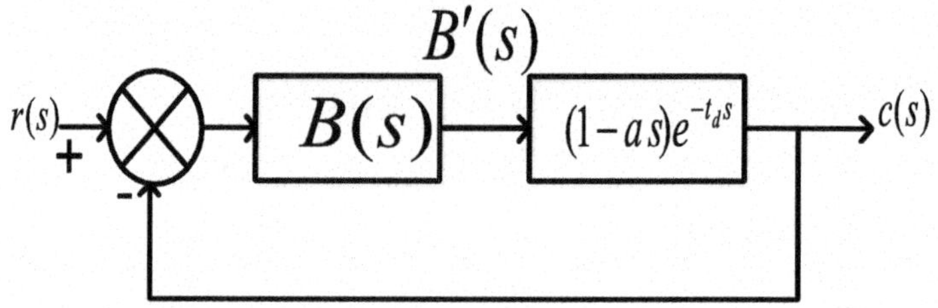

Fig. 7.1 Proposed Bode's ideal transfer function

$$B(s) = \frac{(1 - as)}{\phi s^{\lambda}} e^{-t_d s}, \lambda \in R. \quad (7.1)$$

where a and t_d are the NMP zero and the delay time of the process models.

Method for adaptation

1. Substituting $s = j\omega$ in the Eq. (7.1) then it becomes,

$$B(j\omega) = \frac{(1 - aj\omega)}{\phi (j\omega)^{\lambda}} e^{-t_d j\omega}, \lambda \in R. \quad (7.2)$$

2. Next, by taking the argument of both sides of Eq. (7.2), it transforms into,

$$\angle B(j\omega) = \frac{\angle - \tan^{-1} \omega a - \angle t_d \omega}{\frac{t_d \pi}{2}}. \quad (7.3)$$

3. Subsequently, upon simplifying Eq. (7.3), it can be written as,

$$\lambda = \frac{\pi - PM - \angle - \tan^{-1} \omega b - \angle t_d \omega}{\frac{\pi}{2}}. \quad (7.4)$$

4. After taking the modulus of Eq. (7.4),

$$|B(j\omega)| = \frac{\sqrt{1 + \omega^2 a^2}}{\phi (\omega)^{\lambda}}. \quad (7.5)$$

5. Now adhering, to specification (2), Eq. (7.5) can be expressed as,

$$1 = \frac{\sqrt{1 + \omega^2 a^2}}{\phi (\omega)^{\lambda}}. \quad (7.6)$$

6. After simplifying Eq. (7.6), the time constant can be determined as

$$\phi = \frac{\sqrt{1+\omega^2 a^2}}{(\omega)^{\lambda}}. \tag{7.7}$$

Design of controllers for cascade loop: FOPTD system in the inner loop and First-order plus time delay (FOPDT) system in the outer loop

Inner loop controller design: The inner loop controller is utilized for setpoint tracking and disturbance rejection. The inner loop controller in the IMC framework can be written as,

$$C_s(s) = \frac{f_1(s)}{\tilde{P}^{-}(s)}. \tag{7.8}$$

Here $f_1(s) = \frac{1}{(1+\eta s)^n}$ the low pass filter. n is chosen so that the controller makes it realizable.

Outer loop controller design: The overall plant transfer function description is in Chap. 2, Sect. 2.2. It can be given as,

$$P(s) = H(s)\tilde{P}_1(s). \tag{7.9}$$

From Fig. 7.2, the concept of equivalent system, which is discussed earlier the controller transfer function is achieved as,

$$\frac{C(s)P(s)}{1 + C(s)(I_o(s) + D_o(s))} = B'(s). \tag{7.10}$$

Now the Eq. (7.10) can be written as,

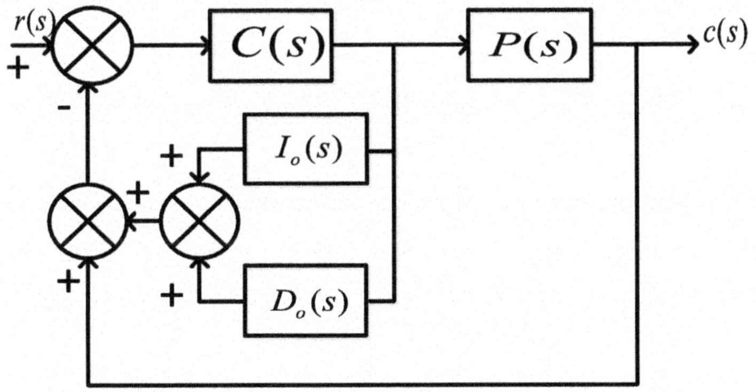

Fig. 7.2 Traditional controller with dead time and inverse response compensator

$$C(s) = \frac{B'(s)}{P(s)}\left[1 + C(s)\left\{\delta s P(s) + \frac{1 - e^{-t_d s}}{e^{-t_d s}} P(s)\right\}\right]. \tag{7.11}$$

Equation (7.11) can be modified as,

$$C(s) = \frac{B'(s)}{P(s)\left[1 - B'(s)\left\{\delta s + \frac{1 - e^{-t_d s}}{e^{-t_d s}}\right\}\right]}. \tag{7.12}$$

Design of controllers for cascade loop: FOPTD system in the inner loop and integer order plus time delay (IOPDT) system in the outer loop

Inner loop controller design: Consider a FOPDT model,

$$\tilde{P}_2(s) = \frac{K_2'}{T_2 s + 1} e^{-t_{d_2} s},$$

where K_2' is the gain, t_{d_2} is the time delay, and T_2 is the time constant.

Inner loop IMC filter.

$$f_1(s) = \frac{1}{\eta s + 1}.$$

where η is the tuning parameter. According to the IMC notion, the inner loop controller transfer function becomes,

$$C_2(s) = \frac{(T_2 s + 1)}{K_2'(\eta s + 1)^n}. \tag{7.13}$$

Here n is chosen in such a way as to make $C_2(s)$ it proper. Moreover, η it can be tuned based on maximum sensitivity,

$$\eta = \frac{-0.7289 M_s + 1.55}{M_s - 1.006} t_d.$$

Here t_d and M_s illustrates the time delay and maximum sensitivity.

Outer loop controller design: Consider an IOPDT system with NMP zero,

$$\tilde{P}_1(s) = \frac{K_1'(1 - as)}{s} e^{-t_{d_1} s},$$

where K_1' is the gain, t_{d_1} is the time delay, and a is the NMP zero.

After utilization Eq. (7.9), the overall plant transfer function becomes,

$$\tilde{P}(s) = \frac{K_1'(1 - as)}{s(s + 1)} e^{-t_d s}, t_d = t_{d_1} + t_{d_2}. \tag{7.14}$$

Now substituting the value $B'(s)$ from Eq. (7.10), $P(s)$ Eq. (7.14), and δ Eq. (7.12). After that, the controller transfer function becomes,

$$C(s) = \frac{\frac{(1-as)}{\phi s^{\lambda}} e^{-t_d s}}{\frac{K_1(1-as)}{s(s+1)} \left[1 - \frac{(1-as)}{\phi s^{\lambda}} e^{-t_d s} (2as + e^{t_d s} - 1)\right]} \tag{7.15}$$

Approximating the delay term via Taylor series expansion and simplifying the Eq. (7.15). It can be modified as,

$$C(s) = \frac{s^{1-\lambda}}{K_1' \begin{bmatrix} \phi + 2at_d s^{3-\lambda} + (2a^2 + 3at_d)s^{2-\lambda} \\ -(2a + t_d)s^{1-\lambda} \end{bmatrix}} (1 + s) \tag{7.16}$$

Design of controllers for cascade loop: FOPTD system in the inner loop and First-order plus time delay (FOPDT) system in the outer loop

Inner loop controller design: The design of the inner loop controller is the same as the previous case,

$$C_2(s) = \frac{(T_2 s + 1)}{K_2'(\eta s + 1)^n}.$$

Outer loop controller design: Consider a FOPDT system with NMP zero,

$$\tilde{P}_1(s) = \frac{K_1'(1 - as)}{T_1 s + 1} e^{-t_{d_1} s}.$$

Now, by considering Eq. (7.9), the outer loop process becomes,

$$\tilde{P}(s) = \frac{K(1 - as)}{(T_2 s + 1)(\phi s + 1)} e^{-t_d s}. \tag{7.17}$$

After utilization Eq. (7.12), the controller transfer function becomes,

$$C(s) = \frac{s^{1-\lambda}}{K\left[\phi + 2ats^{3-\lambda} + (2a^2 + 3at_d)s^{2-\lambda} - (2a + t_d)s^{1-\lambda}\right]}. \left(T_1 T_2 s + (T_1 + T_2) + \frac{1}{s}\right) \tag{7.18}$$

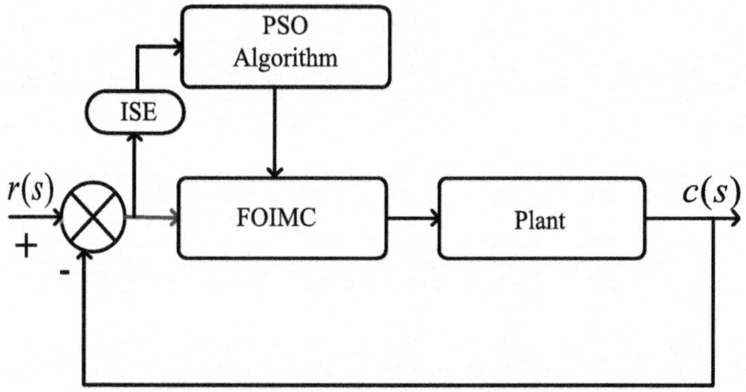

Fig. 7.3 Optimizing controller parameters using PSO algorithm

7.3 Particle Swarm Optimization (PSO)

According to the authors, PSO embodies a simple concept and hypothesis that can be implemented with minimal lines of computer code.

References [5, 6] employed a PSO-based digital FOIMC for tuning Fractional-order controllers. The controller tuning process depicted in Fig. 7.3 employs the PSO algorithm to modify both the filter parameters and the fractional components.

7.4 Case Studies

To evaluate the worth of the suggested tuning approach two benchmark case studies are used.

7.4.1 Case Study I

The prescribed transfer function for the boiler drum level system is given as [3],

$$\tilde{P}_1(s) = \frac{(1 - 0.418s)e^{-0.07s}}{s}, \tilde{P}_2(s) = \frac{0.547e^{-0.03}}{1.06s + 1}.$$

Inner loop Controller: From Eq. (7.13),

$$C_2(s) = \frac{1.06s + 1}{0.547(0.3s + 1)}.$$

Equation (7.14) yields the plant transfer function as,

$$\tilde{P}(s) = \frac{(1 - 0.418s)e^{-0.1s}}{s(0.17s + 1)}. \tag{7.19}$$

Outer loop controller: Eq. (7.15) specifies the recommended controller settings for the outer loop in the fractional IMC context as,

$$C(s) = \frac{s^{0.26}}{1.03 + 0.083s^{2.26}s + 3.34s^{1.26} - 0.936s^{0.26}}(0.17s + 1). \tag{7.20}$$

where the term $\frac{s^{0.26}}{1.03+0.083s^{2.26}s+3.34s^{1.26}-0.936s^{0.26}}$ specifies the fractional IMC filter and the term $(0.17s + 1)$ interprets the PID controller. Here $\eta = 0.17$, $\phi = 1.03$, and $\lambda = 0.74$ with $PM = 80$. The relative analysis of the proposed boiler drum level system method is depicted in Fig. 7.4.

It is observed from Fig. 7.4a, b that the existing controller suffers from both overshoot and undershoot. The goal of the recommended procedure is to minimize peak disturbance and achieve a moderately quick settling time.

In Table 7.1, you can find the IAE and ISE values. The recommended controllers showed convincing effectiveness.

With the outcomes of the boiler drum level system, the detailed procedure of the fractional characteristic polynomial is provided as,

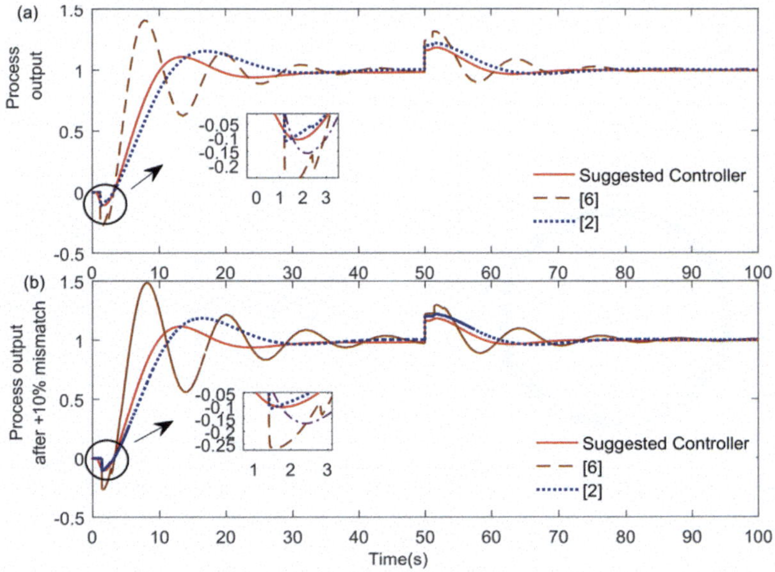

Fig. 7.4 Closed-loop response for boiler drum level system

Table 7.1 Transient response and performance indices for boiler drum level system

Method	Nominal case					+10% mismatch				
	Peak O.S. ratio, %	Peak time, (s)	Rise time, (s)	IAE	ISE	Peak O.S. ratio, %	Peak time (s)	Rise time (s)	IAE	ISE
Suggested controller	1.54	5.81	3.04	0.16	0.02	1.56	5.6	2.99	0.19	0.03
Nagarsheth and Sharma [2]	1.92	1.94	1.03	0.35	0.12	1.95	1.79	0.982	0.37	0.13
Uma et al. [6]	1.73	4.9	2.6	0.24	0.05	1.75	4.90	2.5	0.96	0.92

$$f(s) = 1 + 1.03s^{1.74} + 0.083s^3 + 3.34s^2 - 0.93s.$$

At the same time, assessing the stability of fractional-order polynomials is quite complex.

Thus, mapping $s - plane$ is essential. Arduous $\alpha = s^{\beta}$. After mapping it can be written as,

$$f(\alpha^{1/\beta}) = 1 + 1.03\alpha^{1740} + 0.083\alpha^{3000} + 3.34\alpha^{2000} - 0.93\alpha^{1000}.$$

where $s^{0.001} = \alpha$. The stability condition is assessed on the Riemann surface. Roots located outside indicate system stability, while roots inside indicate system instability (see Fig. 7.5).

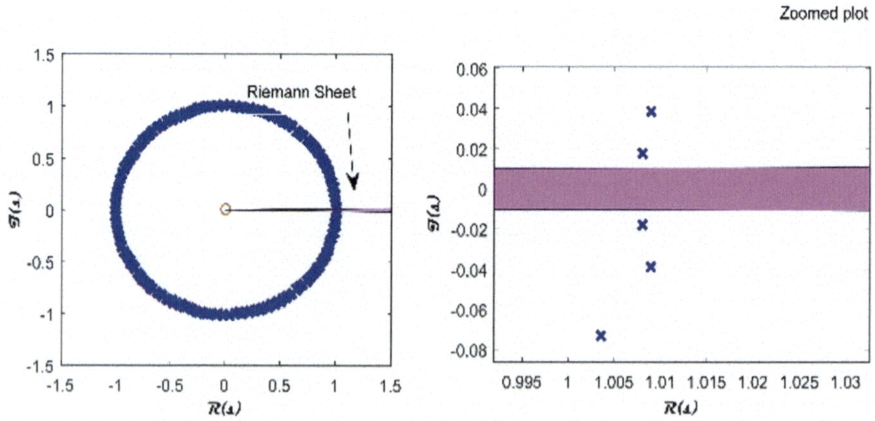

Fig. 7.5 Stability evaluation for boiler drum level system

Assessment of the sensitivity function is achievable [3] (Table 7.2).

$$S(s) = \frac{1}{1 + \tilde{P}(s)C(s)} \tag{7.12}$$

The linear feedback system exhibits different qualitative behaviors as process parameters vary. Figure 7.6 provides a graphical analysis of the robustness of the proposed controller. The recommended controllers exhibit a slight peak in maximum sensitivity under nominal conditions and with a 10% model mismatch. Figure 7.6 also depicts the sensitivity function of the suggested controller alongside two other controllers. Additionally, the PSO algorithm is utilized to compute the ISE value.

The results from Fig. 7.7a, b confirm that the proposed controller, under optimal settings, achieves decreased overshoot, undershoot, and settling time relative to existing controllers.

Remarks 7.1 In this fashion of tuning the NMP zeros and dead-time are reckoned with typical Bode's ideal transfer function. Conversely, existing systems manipulate merely conventional tuning techniques.

Table 7.2 PSO algorithm considerations

Parameter	No. of iterations	C_1	C_2	Swarm size	$W_{max.}$	$W_{min.}$
Value	100	2	1.5	40	1	0.99

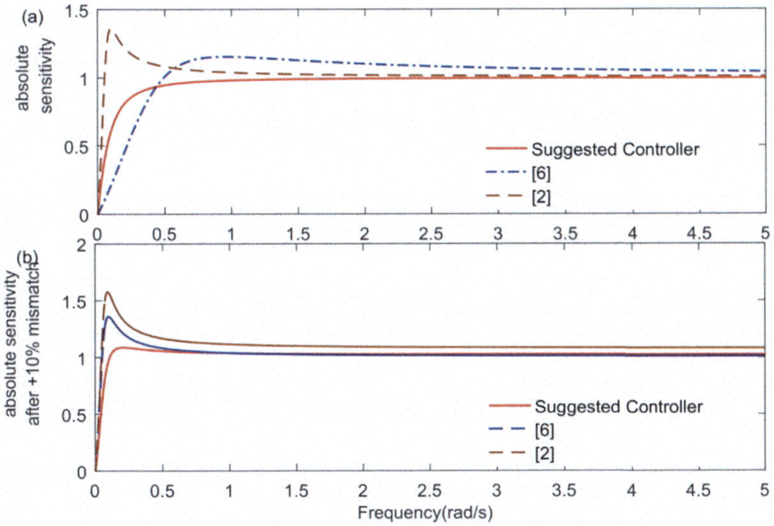

Fig. 7.6 Robustness investigation for boiler drum level system

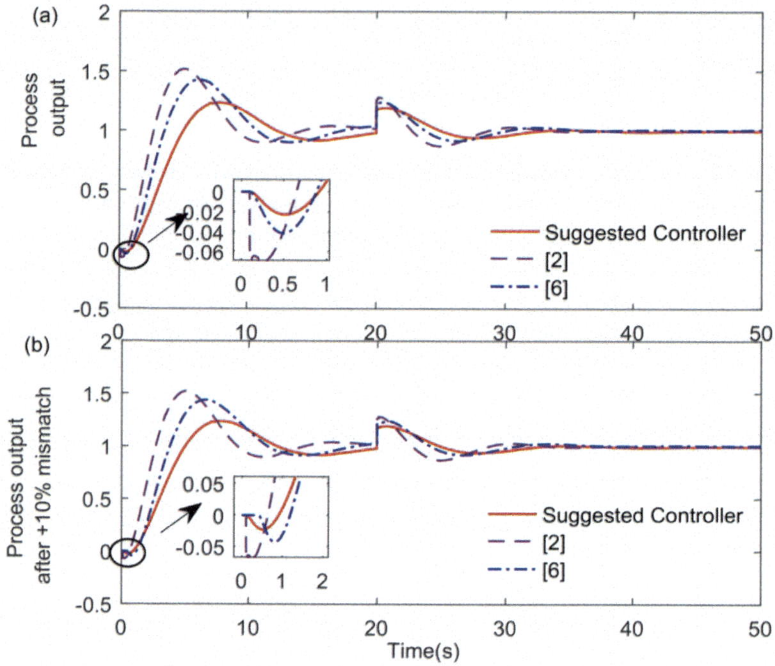

Fig. 7.7 Comparative study of boiler drum level system

7.4.2 Case Study II

The distillation column is characterized by the following transfer function [3],

$$\tilde{P}_1(s) = \frac{(1 - 0.2s)e^{-0.1s}}{s + 1}, \tilde{P}_2(s) = \frac{e^{-0.1s}}{s + 1}.$$

Inner loop controller design: From Eq. (7.13),

$$C_2(s) = \frac{s + 1}{(0.4s + 1)}$$

From Eq. (7.17),

$$\tilde{P}(s) = \frac{(1 - 0.2s)e^{-0.2s}}{(s + 1)(0.4s + 1)}. \quad (7.22)$$

Outer loop controller design: The outer loop transfer function, derived from Eq. (7.18), is

$$C(s) = \frac{s^{0.16}}{1.2 + 0.08s^{2.16} + 0.2s^{1.16} - 0.6s^{0.6}}\left(2 + s + \frac{1}{s}\right). \tag{7.23}$$

where the term $\frac{s^{0.16}}{1.2+0.08s^{2.16}+0.2s^{1.16}-0.6s^{0.6}}$ specifies fractional IMC filter, the term $\left(2 + s + \frac{1}{s}\right)$ reveals PID controller. Figure 7.8a, b postulates step responses with a + 10% discrepancy, the mentioned preparation is characterized by convincible consequences.

Table 7.3 displays the details of transient response and performance measures.

The stability analysis expression for the distillation column is depicted as,

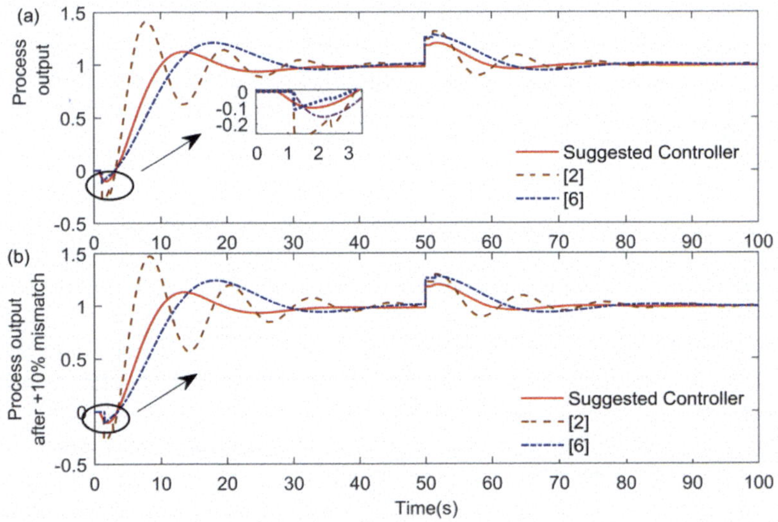

Fig. 7.8 Relative investigation for the distillation column

Table 7.3 Transient response and performance indices for the distillation column

Method	Nominal case					+10%, mismatch				
	Peak O.S. ratio, %	Peak time, (s)	Rise time, (s)	IAE	ISE	Peak O.S. ratio, %	Peak time (s)	Rise time (s)	IAE	ISE
Suggested controller	1.2	3.2	2.3	0.2	0.05	1.34	2.65	2.0	0.2	0.06
Uma et al. [6]	1.5	2.7	1.6	0.5	0.25	1.60	2.36	1.4	0.4	0.06
Nagarsheth and Sharma [2]	1.5	3.3	1.8	0.5	0.3	1.85	2.60	1.3	0.5	0.32

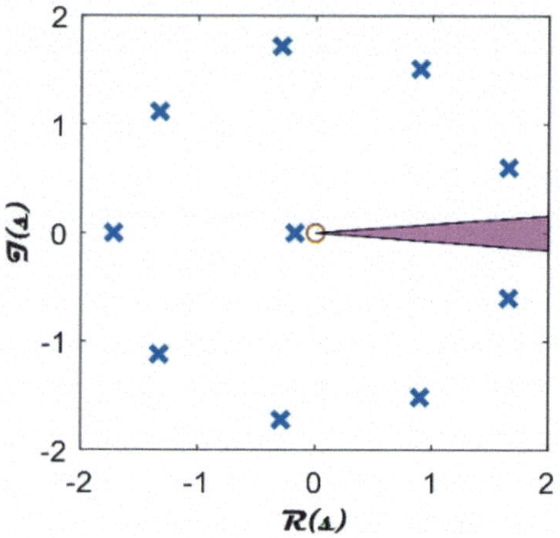

Fig. 7.9 Stability assessment for the distillation column

$$f(s) = 1 + 1.2s^{0.84} + 0.08s^3 + 0.2s^2 - 0.6s.$$

All roots in Fig. 7.9 are located outside the Riemann sheet, verifying the stability of the system.

After using Eq. (7.12) for robustness analysis, Fig. 7.10 demonstrates that the recommended controllers have the minimum and maximum absolute sensitivity function peak.

Figure 7.11 specifies the convergence of the objective function. Figure 7.12a, b confirm better outcomes in both cases (nominal and +10% mismatch).

Remarks 7.2 This tuning approach integrates Non-Minimum Phase (NMP) zeros and dead-time with the conventional Bode's ideal transfer function, contrasting with the current system which employs only conventional tuning technique.

7.5 Summary

This chapter revises the conventional Bode's ideal transfer function by integrating Non-Minimum Phase (NMP) zeros and dead time. It also introduces a suggested tuning technique for fractional-filter parameters, concluding with the use of the PSO algorithm to optimize controller settings.

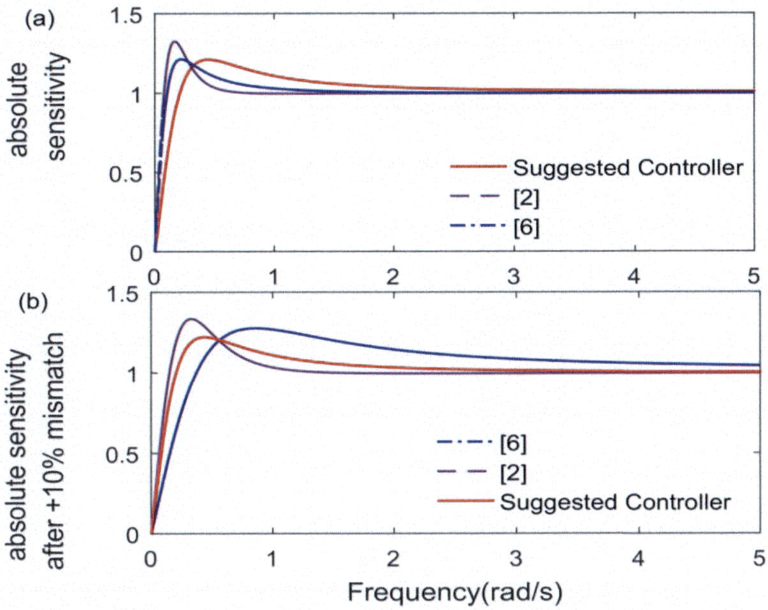

Fig. 7.10 Robustness analysis for the distillation column

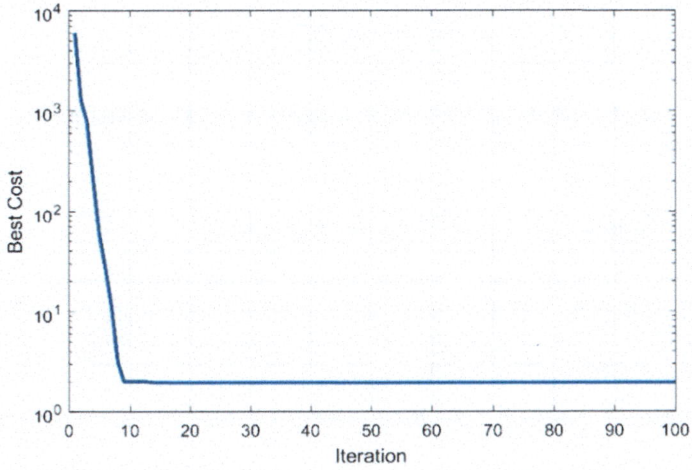

Fig. 7.11 Convergence graph of ISE

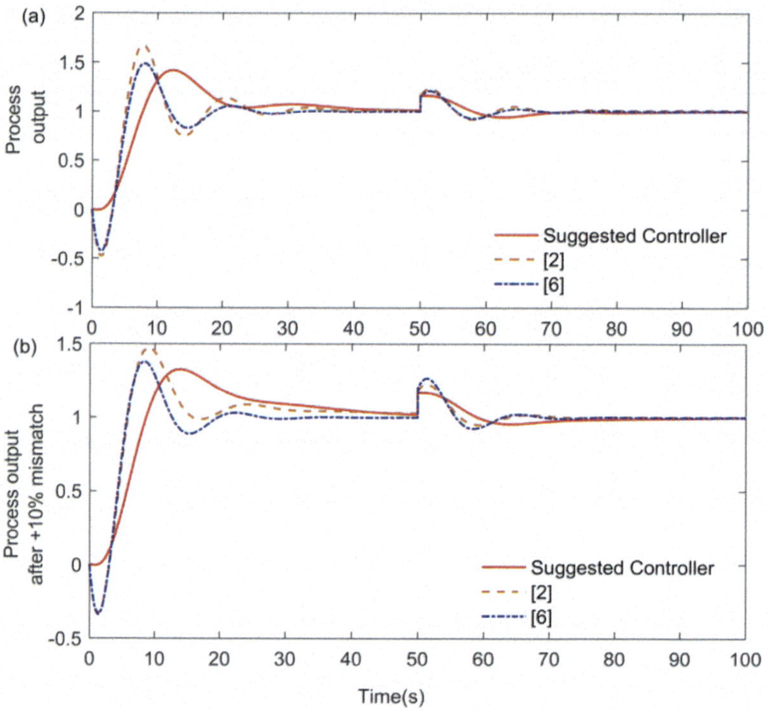

Fig. 7.12 Relative study distillation column using PSO

7.6 Some Informative Views of Chapter in Post-mortem Form

Question 1: What is Bode's Ideal Transfer Function?

Answer: *Bode's ideal transfer function refers to a theoretical representation used in control systems and signal processing. It typically consists of a combination of simple transfer functions like integrators and differentiators to analyze the frequency response characteristics of a system.*

Question 2: How do you paradigm Bode plots using Bode's ideal transfer function?

Answer: *To construct Bode plots, you evaluate the transfer function at different frequencies (ω) ranging from very low to very high values. For each frequency, you determine the magnitude (in dB) and phase shift (in degrees) of the transfer function.*

Question 3: How does Bode's ideal transfer function relate to the Nyquist stability criterion?

Answer: *Bode's ideal transfer function can be used to analyze stability using phase and gain margins. These margins are related to the positions of poles and zeros of the transfer function relative to the imaginary axis, which is also a key concept in the Nyquist stability criterion.*

References

1. Ranganayakulu R, Rao AS, Bhasakar B (2020) Analytical design of fractional IMC filter–PID control strategy for performance enhancement of cascade control systems. Int J Syst Sci 51(10):1699–1713
2. Nagarsheth SH, Sharma SN (2020) Control of non-minimum phase system with dead time: a fractional system viewpoint. Int J Syst Sci 51(11):1905–1928
3. Yumuk E, Güzelkaya M, Eskin I (2019) Analytical fractional PID controller design based on Bode's ideal transfer function plus time delay. ISA Trans 91:196–206
4. Khalifa B, Abdelfateh C (2017) Optimal tuning of fractional order A controller using particle swarm optimization algorithm. IFAC-OnLine 50(1):8084–8089
5. Su S, Zhai Z, Wang C, Ding K (2021) Improved fractional-order PSO for PID tuning. Int J Pattern Recogn Artif Intell 35(6):2159016
6. Uma S, Chidambaram M, Rao AS, Yoo CK (2010) Enhanced control of integrating cascade processes with time delays using modified Smith predictor. Chem Eng Sci 65(3):1065–1075
7. Shamsuzzoha M, Lee M (2008) Design of advanced PID controller for enhanced disturbance rejection of second-order processes with time delay. AIChE J 54(6):1526–1536
8. Joshi D, Deb D, Giri AK (2024) MRAS disturbance observer-based sensorless field-oriented backstepping control of BLDC motor drive. In: Electrical engineering. Springer, Berlin. https://doi.org/10.1007/s00202-024-02378
9. Kundu S, Singh M, Giri AK (2022) Implementation of in-loop filter based DSC-FLL control approach for power quality improvement in isolated microgrid. In: Proceedings of 2022 IEEE 31st international symposium on industrial electronics (ISIE). USA. https://doi.org/10.1109/ISIE51582.2022.9831712
10. Giri AK, Arya SR, Maurya R (2020) Hybrid order generalized integrator based control for VSC to improve the PMSG operation in isolated mode. In: The proceedings of 2020 first international conference on power, control and computing technologies (ICPC2T). India. https://doi.org/10.1109/ICPC2T48082.2020.9071498
11. Kundu S, Singh M, Giri AK (2024) Synchronization and control of WECS-SPV-BSS-based distributed generation system using ICCF-PLL control approach. Electric Power Syst Res 226:109919
12. Giri AK, Arya SR, Guerrero JM, Kumar S, Power quality: infrastructures and control. Springer Nature, Singapore. ISBN: 978-981-19-7956-9
13. Giri AK, Arya SR, Vinnikov D, Distributed energy systems design, modeling, and control, vol 1. CRC Press, pp 1–308. ISBN: 9781032134246
14. Arya SR, Patel MM, Alam SJ, Srikakolapu J, Giri AK, Chitti Babu B (2021) lassical control algorithms for permanent magnet synchronous generator driven by diesel engine for power quality. Int J Circuit Theory Appl 49(3):576–601
15. Kundu S, Giri AK, Singh M, Kadiyan S, Adaptive control strategy for isolated renewable energy-based generation system with intermittency. In: Electrical engineering. Springer, Berlin. https://doi.org/10.1007/s00202-024-02454-0
16. Meena DC, Singh M, Giri AK (2021) Leaky-momentum control algorithm for voltage and frequency control of three-phase SEIG feeding isolated load. J Eng Res ICARI Spec 109:120
17. Meena DC, Singh M, Giri AK (2021) A modified nlms control algorithm for coordinated operation in three-phase wind-energy conversion system. Int J Renew Energy Res (IJRER) 11(4):1621–1629

18. Dalwadi N, Deb D, Ozana S (2023) Performance evaluation of RW-quadrotor and Bi-quadrotor for payload delivery. IET Intel Transp Syst 17(11):2221–2236
19. Dalwadi N, Deb D, Ozana S, Adaptive hybrid control of quadrotor drones. 461:1–174. ISBN: 978-981-19-9743-3

Chapter 8
Conclusion and Future Scope

8.1 Summary of Findings

1. **Series Cascade Control in IMC Framework**:
    - Approach: Designed an outer loop controller for a series cascade scheme in the IMC framework using three approximations to integrate inverse response and dead time compensators.
    - Results: Applied effectively to two practical problems, with robustness assessed through sensitivity analysis.
2. **Improved Parallel Cascade Control**:
    - Approach: Proposed a novel parallel cascade control method for non-minimum phase systems with dead time.
    - Results: Enhanced performance indices demonstrated through numerical simulations.
3. **Revised Outer Loop Controller Design**:
    - Approach: Enhanced the outer loop controller by integrating an inverse response compensator and fractional filter, optimized via Particle Swarm Optimization (PSO).
    - Results: Limited undershooting and improved performance without additional controllers or setpoint filters, validated in practical scenarios.
4. **Modified Bode's Ideal Transfer Function**:
    - Approach: Adjusted Bode's ideal transfer function for NMP zeros and time delays, designed the controller within the IMC framework, and tuned the fractional filter using PSO.
    - Results: Confirmed the effectiveness of inverse response compensation and improved control robustness, with a sensitivity function closer to one and elimination of Smith predictor for small dead time.

M. Yadav et al., *Control Strategies for Non-minimum Phase Systems*,
Studies in Infrastructure and Control, https://doi.org/10.1007/978-981-96-2351-8_8

8.2 Future Scope

Consider embedding Linear Active Disturbance Rejection Control (LADRC) into the proposed control methodology to improve disturbance rejection.

The manufacturer's authorised representative in the EU is Springer Nature Customer Service Centre GmbH, Europaplatz 3, 69115 Heidelberg, Germany. If you have any concerns regarding our products, please contact ProductSafety@springernature.com

Printed and bound by CPI Group (UK) Ltd, Croydon, CR0 4YY
01/10/2026
02225945-0014